D0717456

Elements
of
Mathematics

Elements
of
Mathematics

Under the Editorship of the Late Carl B. Allendoerfer

PRENTICE-HALL, INC.
ENGLEWOOD CLIFFS, NEW JERSEY

James W. Armstrong

University of Illinois, Urbana

Second Edition

Macmillan Publishing Co., Inc.
New York
Collier Macmillan Publishers
London

NORTHWEST MISSOURI STATE
UNIVERSITY LIBRARY
MARYVILLE, MISSOURI 64468

Copyright © 1976, James W. Armstrong

Printed in the United States of America

All rights reserved. No part of this book may be reproduced or transmitted in any form or by any means, electronic or mechanical, including photocopying, recording, or any information storage and retrieval system, without permission in writing from the Publisher.

Earlier edition copyright © 1970 by James W. Armstrong.

Macmillan Publishing Co., Inc.
866 Third Avenue, New York, New York 10022

Collier Macmillan Canada, Ltd.

Library of Congress Cataloging in Publication Data

Armstrong, James W
  Elements of mathematics.

  Includes index.
  1. Mathematics—1961–   I. Title.
QA39.A73   1976   510   74-24367
  ISBN 0-02-303910-8

Printing: 1 2 3 4 5 6 7 8   Year: 6 7 8 9 0 1 2

510
A73e
1976

Preface

In the coming chapters we shall discuss some aspects of eight or nine of the most important and well-known branches of undergraduate mathematics. Every effort has been made to bypass the usual computational aspects of the topics covered so that many chapters involve no computation at all. In those few instances where a minimum of computation is absolutely necessary, we have included sufficient preliminary work in techniques. The student is not required or asked to bring more into his classroom than a willingness to look at some old and some new ideas with an unprejudiced mind.

The chapters are organized so that the text may be used in a variety of courses, although the primary purpose of the book is to present topics for what is often called a "liberal arts mathematics" course. Much effort has gone into designing chapters and sections that are sufficiently independent so that an instructor can easily skip those sections and chapters dealing with material not pertinent to his course. These chapters and sections are designated in the text by a dagger. Most chapter interconnections (aside from the obvious ones) appear only in the exercises, and the teacher's manual contains a finely detailed identification of these interconnections. The manual also contains carefully formulated estimates of the number of 50-minute sessions that might be needed to cover each section. This material is designed to assist the instructor in molding the contents of the text into the form best suited for his class.

The teacher will find a number of starred exercises in the text. The star does *not* indicate an exercise of more than average difficulty. It indicates that the exercise deals with an idea introduced in a previous exercise or with an idea that has not been explicitly discussed in the body of the text. Starred exercises of the latter type may be used as topics for additional class discussion.

I wish to acknowledge the assistance of the late Professor Carl B. Allendoerfer and of Professor Ronald E. Walpole, who reviewed the statistics

JUL 11 1980

chapter. Shirley Wilson and Debra Romack assisted most ably in the preparation of the answers. Mostly, however, I must acknowledge the assistance of my wife, Dianna, who has worked alongside me throughout this revision and whose insight has had a measurable bearing upon the improvements that appear here. In a very real sense she is a co-author.

J. W. A.

Contents

2

Mathematical Systems *43*

3

The System of Whole Numbers *71*

4

The Theory of Sets *99*

5

Two Geometries   *127*

6

Extending the Concept of Number   *159*

7

Analytic Geometry   *189*

Elements
of
Mathematics

Flow Charting

The use of computers has produced a method of representing mathematical and other procedures pictorially by means of diagrams called **flow charts.** In this preliminary chapter we shall discuss flow charting. We shall use these charts in later chapters to augment and illustrate the discussions in the body of the text.

0.1 Communication Between Men and Machines

The fact that communication is possible between men and machines ought not to surprise us, for we all communicate with machines every day. We can, for example, talk with a radio. We can't say to the radio, "Turn yourself on and tune in on wavelength 99.5" and expect the radio to do it, because the radio can't understand our English language words. What we have to do is to translate these English language words into "words" that are comprehensible to the radio. We have to translate from the language of the user to the language of the machine.

The author has a radio so constructed by its makers that the instruction "Turn yourself on" is rendered into radio language by flipping a certain switch to the left. Flipping that switch is the translation into radio language of the English language instruction "Turn yourself on." The radio language equivalent of the

English language instruction "Turn to wavelength 99.5" is given to the radio by pushing a button marked "B." Thus, if one person stands in front of this radio shouting "Turn yourself on," and another person walks up to the radio and flips the switch, then we would say that the only difference between these two people lies in the fact that the first person doesn't understand how to give instructions to that radio—he does not understand radio talk—but that the second person does.

Men can talk with automobiles as well. We communicate an instruction to an automobile when we turn the ignition key. The instruction that we are communicating in this way is the instruction that is rendered in English as "Turn yourself on." The instruction that we give to an automobile when we put our foot on the brake is rendered in English by the word "Stop."

Both the radio and the automobile were designed by their builders to respond to certain selected kinds of instructions. Generally, in such machines as radios and cars, for each instruction that the machine is designed to recognize, there is a special device built into the machine that will communicate the human operator's desire to have that instruction carried out.

Computers are also constructed according to this general principle, but the difference is profound because computers are designed to follow a great many more instructions than radios and automobiles. Because its capabilities go far beyond those of the radio, a computer must be able to receive a great many more instructions; and this makes it much more difficult to communicate with a computer. It would not be feasible to install a separate switch for each and every instruction that was to be followed because computers must be able to respond to thousands of instructions.

A new way is needed, therefore, to communicate instructions to a computer. The method that has been developed is to construct the computer with its own special numerical language built in. This numerical language consists only of numbers and is so complicated that only the computer itself can possibly understand it. So the computer language is impossible for humans to use. On the other hand, English is impossible for the computer to use. What we need is someone to translate the English language instructions of the user into the numerical language of the computer.

The person who translates the English language instructions into a language computers can understand is called a **computer programmer.** Programmers take English language instructions and put them into special programmer language (FORTRAN, COBOL, or BASIC, among others), which is much nearer to the computer's own numerical language but still understandable to people with the requisite training. Moreover, their language is sufficiently close to the numerical language of the computer so that the computer itself is able to translate the programmer's language into the numerical computer language. Hence translating English instructions into computer language is a two-step process. In the first step the programmer translates the English instructions into the programmer's special language. The second step is performed by the computer itself when it translates the programmer's special language into its own numerical language.

The programmer's job is in fact a very difficult one. If the set of instructions is very long (and most often they are), the job of the programmer could take weeks of very hard labor. The first step is for the programmer to become familiar with the instructions the user has given him. Then he constructs a kind of first approximation to the programmer's language translation that he is seeking. He lays out his English language instructions in a precise way by means of what we call a **flow chart.** Flow charts enable the programmer to organize even a very complicated set of instructions into a logical display that shows fairly clearly what the computer will have to do in order to carry out the instructions. The programmer then works from the flow chart to construct the final product of his art, the programmer's language instructions. We shall discuss flow charting in the next section and look at some examples of mathematical instructions and nonmathematical instructions organized by means of flow charts.

Exercises 0.1

1. Make a list of some of the instructions that a human may communicate to each of the following types of machines:

   (a) A table lamp.   (b) A pencil. (Careful!)
   (c) A typewriter.   (d) Your left hand.
   (e) A telephone.   (f) A desk stapler.
   (g) An elevator.   (h) An automobile.

2. Can you think of some machines that are designed to be able to read words, numbers, or symbols in some way, even some very limited way? Make a list and describe briefly what they are designed to read and what they are not designed to read.

3. Object recognition is one of the new applications of computers. For example, computers can be linked to television cameras and taught to tell the difference between a circle and a square or between a square and a rectangle. However, this is not too spectacular a feat. We are all familiar with machines that have been "taught" to discriminate between certain very special types of objects. Make a list of such machines and the objects that they are designed to recognize. (*Suggestion:* A dollar bill changer.)

0.2   Constructing Flow Charts

In this section we learn to construct flow charts for mathematical and non-mathematical operations.

Let us consider the job of an assembly line worker whose job it is between

8 A.M. and noon and between 1 P.M. and 5 P.M. to insert one baseball after another into its own individual box, close up the boxes, and put them on a conveyor belt, whereupon they are carried away. (We don't care where.) The supervisor of such a worker might very well train new workers with the following simple instructions: "Look, you sit here in this chair. To your left will always be a crate of baseballs and to your right will always be a crate of boxes. You are to pick up a baseball and a box, put that baseball into that little box, close up the top of the box, and then put the box onto this conveyor belt that is passing along right here in front of you. You do this starting at 8 A.M. and stopping at 5 P.M. You get from noon to 1 P.M. for lunch. OK, get to work."

For most people these instructions would be perfectly satisfactory, for the job is not complicated. But let us consider these instructions carefully. Perhaps of most interest is the part of these instructions that involves the actual job of placing the balls into the boxes and closing up the tops and then placing these boxes on the conveyor belt. We could refer to placing a ball into a box and closing the top of the box as "packing a box" and most everyone would understand. We can describe this portion of the job pictorially as shown in Figure 0.1.

This flow chart describes the actual individual piece of work that the worker does all day long many times over. Next, let us add to this flow chart the information that the worker is to repeat this piece of work over and over again. We do this by introducing a **loop** into the flow chart, as shown in Figure 0.2.

The only thing wrong with this flow chart is that now we have the worker working forever. We have not inserted instructions that will tell the worker to stop under certain conditions. Let us correct this flow chart for the morning shift. See Figure 0.3.

What we have added to Figure 0.3 is a **decision step.** We now have a flow chart that instructs the worker to stop by noon but to keep packing until then. Our flow chart does not yet tell the worker when to begin work so that we need

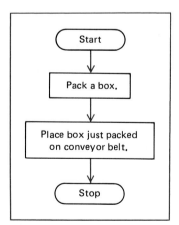

Figure 0.1

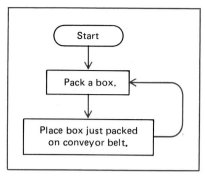

Figure 0.2

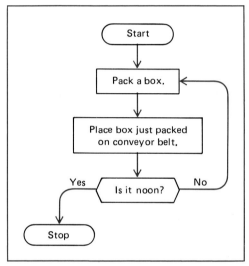

Figure 0.3

to add more instructions at the top of the chart. When should the worker start work? At 8 A.M., according to the supervisor. We must add another decision step, as shown in Figure 0.4, which will tell the worker whether or not it is time to begin work.

Notice that in order to tell the worker when to start in the morning, we have inserted a loop. If you have ever been around workers waiting for 8 A.M. before getting to work, you will have seen people doing just exactly what is described here. The loop that tells the worker when to stop for lunch might be called the "clock-watching loop."

Now, this chart is the morning chart. There is also an afternoon chart that is exactly the same except that 8 A.M. is replaced by 1 P.M. and noon is replaced

---

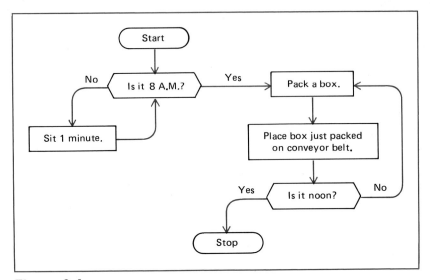

Figure 0.4

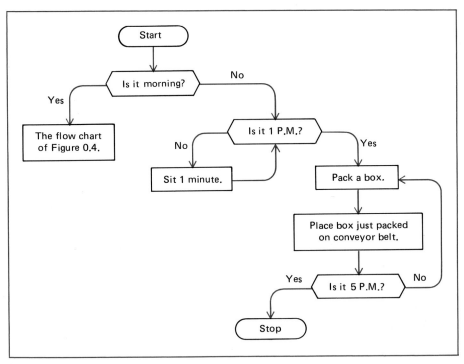

Figure 0.5

by 5 P.M. So, the total day's instructions consist of two *subcharts,* which are shown in Figure 0.5.

Looking at Figure 0.5, one could correctly argue that the chart is really incomplete. For example, the chart does not tell the worker precisely how to pack the box. Which hand should he use to pick up the baseball? Most boxes have three flaps to be closed; which should be closed first and which second? How exactly does one place a packed box on the conveyor belt? Do you heave it (after all these are baseballs) or put it down very gently? What happens if the worker who is supposed to supply you with crates of boxes gets behind and at some stage you don't have any more boxes? One could obviously go on forever (almost) with such questions and such legitimate criticisms of the flow chart we have constructed. In fact, if we were programming a computer to do this job, then we would have to answer all these questions, and more besides. But enough of this kind of talk. We are interested in programming human beings to do things and to understand things. We are not now in the business of instructing computers. We can say to a human being "Put the ball into the box and close up the box," and the human being will have a clear and sufficiently precise understanding of what needs to be done to do the job. The computer wouldn't, but that is not our problem—it is the computer programmer's problem.

The remainder of this chapter consists of a number of examples of the flow charting of familiar mathematical operations.

Example 1.   A very simple mathematical operation is that of finding the average of two whole numbers. The operation is performed by summing the two numbers and then dividing this sum by 2. The resulting quotient is the average of the two numbers. See Figure 0.6. The instructions for finding the average are easily flow charted because there are no loops involved and there are no decisions to be made.

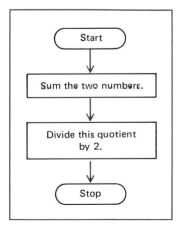

Figure 0.6

Example 2. The determining of whether or not a number x is even. See Figure 0.7.

Example 3. The single, most difficult computational operation that is taught in elementary school is the long division process. This process is introduced as early as the second or third grade, but students in the ninth and tenth grades often have difficulty in working even fairly routine problems. The process that you use to find quotients and remainders in long division problems is the one that is hardest to describe. This is the process that starts out, for example,

$$17\overline{)370}$$

and begins by your estimating the number of 17's contained in 37. To write out a flow chart for this algorithm is rather difficult; but after you have more experience with flow charting, you may be interested in trying your hand at it.

The easier process for finding long division quotients and remainders (the one that is easier to explain, although seldom the easiest to use in practice) involves repeated subtraction. That is, to long divide 370 by 17 you may simply begin subtracting 17 from 370 and keep subtracting 17's until you finally obtain a difference that is less than 17. The total number of times you will have subtracted 17 is the quotient to the long division, and the difference that you ended up with when you stopped is the long division remainder.

Figure 0.8 contains a flow chart that displays this process in a reasonably careful way. Note that because the process will, in general, require repeated subtractions, it will have to contain a loop. Because the repeated subtractions do not repeat forever, there must be a decision at the end of the loop that will tell us whether to go into another loop or whether to stop looping and terminate the chart. The

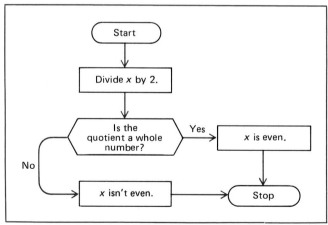

Figure 0.7

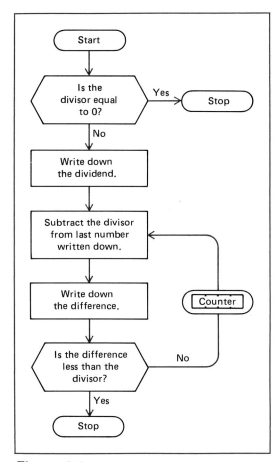

Figure 0.8

counter is a device that tells us how many times we have gone through the loop. For our purposes, counters will always be set at zero initially. Each time we go through the loop, one more will be added to the number in the counter. The quotient is 1 greater than the number registered on the counter, and the remainder is the last difference that was written down.

Example 4. We often must "round off" fractional numbers to whole numbers. For example, if three cans of tomato paste are sold for \$1, then you will have to pay 34 cents for a single can. Of course, the exact cost for a single can should be $33\frac{1}{3}$ cents, but the store will always round such fractional prices upward to the next whole penny. The flow chart in Figure 0.9 describes the process by means of which the selling price of a single item is determined if n items are sold for C cents.

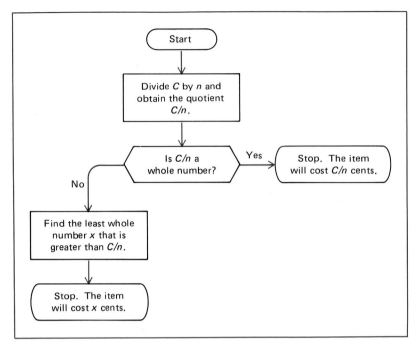

Figure 0.9

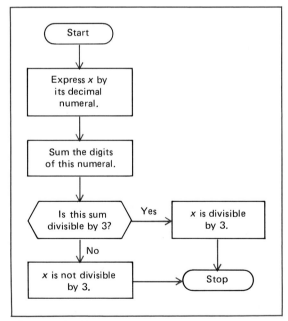

Figure 0.10

Example 5. There is a fairly simple test to tell whether a whole number is evenly divisible by 3 (that is, is divisible with no remainder). The test is described in Figure 0.10 by means of a flow chart. For example, the number 123123 is divisible by 3 because the sum of the digits of this number is 12 $(1 + 2 + 3 + 1 + 2 + 3 = 12)$, which is divisible by 3. Another example is the number 104567898. The sum of the digits of this number is 48, which is divisible by 3. Hence the original number is divisible by 3.

Example 6. How do you find the largest number in a group of numbers? For example, consider the numbers 15, 8, 13, 5, 12, and 7. Probably, you scan this arrangement of numbers moving your eyes from left to right. Whenever you come to a new number that is less than the largest of the numbers you had previously scanned, you ignore the new number. When you come to a new number that is larger than the largest of the numbers previously scanned, you remember the new number and discard the previous largest number. This sort of verbal description is fairly complicated and is slightly difficult to write and to understand. But this can be put into flow chart form, and then the instructions become more routine and machinelike. See Figure 0.11. The last number written down will be the largest.

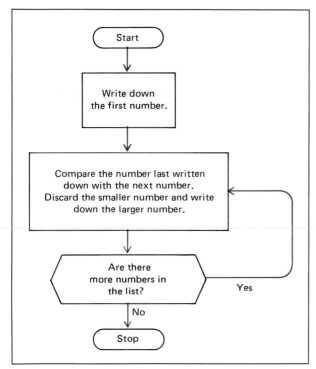

Figure 0.11

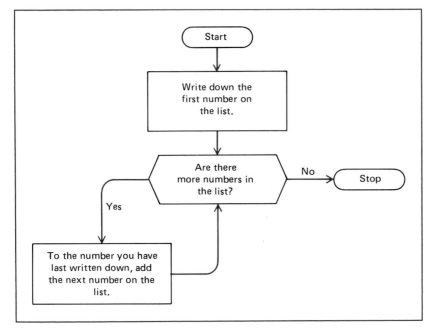

Figure 0.12

Example 7. Flow charts for the usual addition process are fairly complicated (because of the carrying procedure), but if we assume that we already know how to add *two* whole numbers, then we can construct a flow chart describing the procedure for adding the numbers in any list regardless of it length. See Figure 0.12. The last number written down is the sum of all the numbers on the list.

We shall employ flow charts throughout the text. There are many exercises calling for you to construct flow charts. Pay attention to these exercises because they can be used for self-testing. If you can construct a flow chart for an operation, then the chances are very good that you understand the operation rather well. If you have trouble with the flow chart, then it may be that you only partially understand the operation. You may not always construct the same flow charts as those given in the text or in the answers. Each problem may be correctly answered by two or more flow charts, and, in general, one flow chart is as good as another for our purposes.

Exercises 0.2

1. Construct flow charts for performing the following operations:
   (a) Determining whether or not a number is odd.
   (b) Finding the average of three numbers.

(c) Finding the largest of three numbers. Do not use a loop.

(d) Arranging three whole numbers in increasing size.

(e) There is another common rounding off process. If the fractional part of the cost of an item is greater than or equal to one half, then round the cost upward to the next penny. But, if the fractional part of the cost is less than one half, reduce the cost to the next lowest penny. For example, $45\frac{1}{3}$ cents would be rounded off to 45 cents, whereas both $45\frac{1}{2}$ cents and $45\frac{2}{3}$ cents would be rounded off to 46. Describe this process by means of a flow chart.

2. Construct flow charts for performing the following tasks.

(a) Starting an automobile.

(b) Trying to open a lock with a selection of three keys when you don't know which key is the right one. Do not use a loop.

(c) Same as (b), but use a loop.

(d) Trying to open a lock with a selection of any number of keys when you don't know which key is the right one. You will have to use a loop here.

(e) Crossing an intersection at a stop light.

(f) Frying bacon.

(g) Memorizing a poem.

(h) Turning on and adjusting the color for channel 3 on your TV.

(i) Changing an automobile tire.

3. Is it possible to have a flow chart with a loop that is not followed by a decision step?

4. Construct an example of a set of instructions whose flow chart has two loops, one following the other. (There will have to be a decision step between the two loops.)

Logical Foundations

It is appropriate that we should begin our study of the underlying ideas of modern mathematics with an examination of the logical foundation for those ideas. To do this, we must first beat a quick retreat back to the time of the ancient Greeks and then work our way forward again to about our own time. We shall begin with a discussion of the axiomatic method and conclude with a look at elementary symbolic logic.

1.1 Axiomatic Mathematics, I

The characteristic of modern mathematics that so dramatically differentiates it from the other sciences is that mathematics is presented as an axiomatic study, whereas the other sciences are axiomatic only to the extent that they utilize mathematics. Let us begin by a brief examination of axiomatic mathematics.

Exactly when, where, and by whom axiomatic or deductive reasoning was first used in mathematics is unknown, although there are conjectures that it might have been a Greek of the fifth century B.C. named Thales of Miletus. It is likely that until this time mathematics did not exist as we know it today but instead consisted of unrelated and empirically derived ideas about numbers and simple geometric objects, such as circles and right triangles. The first specific evidence of an axiomatic treatment of a part of mathematics is contained in an elementary textbook concerned with geometry, algebra, and number theory called *Elements* written by a Greek, Euclid of Alexandria, about 300 B.C. Probably, you have heard of Euclid only as a geometer, although the real significance of his textbook lies not so much in the geometry it contains as in the way this geometry was presented. Euclid was, as far as we know, the first writer of mathematical literature to employ the axiomatic or deductive method of reasoning, and it is for this reason that he is famous.

Axiomatic or deductive reasoning is the key to modern mathematics. The key to **deductive reasoning** is the idea that the truth of a statement must be shown to follow logically from the truth of other statements, which have already been shown to be true by this method. Loosely speaking, the truth of a new statement must be deduced from the truth of old statements.

This is in contrast to the other method of reasoning called **inductive reasoning.** When we establish the "truth" of a statement by making a generalization based upon a limited number of observations or experiments, we are using inductive reasoning. But inductive reasoning has one great fault: We have no guarantee that an event that has taken place in the same way each time we observed it will continue to take place in that same way in the future. Also, it may be that our observation was faulty or that we did not correctly interpret what we observed. It is because the theories of physics, for example, are based for the most part on inductive reasoning that every so often physical theories must be revised so as to agree with new observations made since the last time the theory was formulated. Generally speaking, inductive reasoning plays a great role in the nonmathematical sciences and plays relatively little role in mathematics. Deductive reasoning, on the other hand, plays a relatively small role in the nonmathematical sciences but plays a great role in mathematics.

But insistence upon deductive reasoning in mathematics does present a problem. If the truth of each statement must be based upon the truth of previously proved statements, how was the *first* statement proved true? When trying to prove the first statement, there are no previously proved statements available upon which to base the truth of this first statement. The answer is that it could not have been proved true at all, and so it can only have been assumed to be true. Thus every study that involves deductive reasoning must necessarily involve a number of statements that are accepted as being true without proof. These statements are called **postulates** or **axioms.** The axioms are the statements that are used to prime the deductive pump. After the pump has been primed, it can begin producing other true statements. When Euclid wrote his *Elements,* he started with these five postulates:

1. A line can be drawn from any point to any other point.
2. A segment can be extended continuously to a line.
3. A circle can be drawn with any point as center and any segment as radius.

4. All right angles are equal.
5. Given a line l and a point P not belonging to l, there can be drawn through P exactly one line that is parallel to l.[1] (See Figure 1.1.)

On the basis of these five postulates (together with additional postulates that were introduced as needed), Euclid deductively built up a geometry that we call Euclidean geometry, with which you are familiar to a greater or lesser degree. Each statement called true in this geometry is either one of the postulates (which are assumed to be true) or one of the proved statements (which are variously called theorems, propositions, lemmas, or corollaries).

Now, some discretion must be exercised in the choice of the statements to be called axioms. Among a number of other criteria, we want the list of axioms to be as concise as possible and we want each axiom to express as primitive an idea as possible. In other words, we do not want to assume more than we really have to assume. But most important of all, we want the collection of axioms to be **consistent.** By this we mean that we do not want the axioms to give rise to logical contradictions. For example, we would not want to choose axioms that would eventually give rise to both the statements, "The sum of the angles of a triangle is 180°" and "The sum of the angles of a triangle is less than 180°." Euclid's geometry is consistent in the sense that the axioms do not give rise to logical contradictions.

Example 1. Arithmetic can be organized as a deductive system. Suppose we were to try to do this by using as a part of our list of axioms these four axioms:

(a) $2 + 3 = 1$.
(b) $1 + 2 = 2$.
(c) $(2 + 3) + (1 + 2) = 8$.
(d) Equals added to equals are equal.

The resulting system would be *inconsistent.* Why?

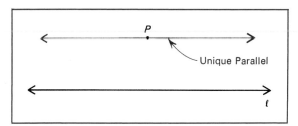

Figure 1.1

[1] By a parallel to l through P we mean a line passing through P that does not intersect l. Incidentally, this statement is a simplified version of Euclid's original, more complicated statement.

Solution: The fact that the first two axioms do not agree with our usual understanding of the arithmetic of numbers does not show that the system is inconsistent. Remember, these statements are being taken as axioms and hence are true by assumption. The reason they are inconsistent is that, using axiom (d), axioms (a) and (b) can be combined into the proved statement

$$(2 + 3) + (1 + 2) = 1 + 2 = 2$$

and this *theorem* is in contradiction to axiom (c), which says that the sum equals 8. The fact that axiom (c) contradicts a statement provable in the system shows that the system is inconsistent.

Euclid's *Elements* is remarkable for the great advance it made in understanding the way that deductive logic is involved in mathematics, but nevertheless there were serious errors in Euclid's logic. One of these was his failure to recognize the fact that he could not hope to give a definition to each and every concept with which he dealt. If you will reflect for a moment upon the fact that definitions of new concepts are phrased in terms of previously defined concepts, then you will see that just as we needed axioms to get the statement-proving process started, we need undefined concepts to get the concept-defining process started. Because we cannot hope to find a definition for the first concept in terms of previously defined concepts, this first concept (along with a few others) must be taken without definition. Euclid missed this fact and attempted to define every concept. As a consequence many of his definitions lack any real power as definitions. For example, he defined a line as "that which is breadthless length" without having previously defined what he meant by "length" and by "breadth." Such definitions are of no value whatsoever.

Finally, by a **deductive** (or axiomatic) **system** we mean a part of mathematics that has been organized along the following lines. First of all, we have selected a number of concepts that we feel are very primitive and that we agree to accept without definition. These are the undefined concepts of the system. Second, we have selected some statements concerning these undefined concepts that we feel express very primitive truths about the undefined concepts and that we are going to accept without proof. These are the axioms of the system. Then, using the undefined concepts and axioms we can begin the process of defining new concepts in terms of the undefined concepts and establishing the truth of new statements about these concepts on the basis of the axioms. The resulting deductive system, therefore, consists of four basic parts: undefined concepts, axioms, defined concepts, and theorems. One of the goals of mathematics is to present every part of mathematics in the form of such a deductive system.

Exercises 1.1 _____

1. Sherlock Holmes concluded that the man lurking around his hotel was a coal miner, after hearing his cough and seeing his blackened fingernails. Was Watson correct in exclaiming, "Brilliant deduction, Holmes!"?

2. You are told by John's roommate that John is either in the lounge watching television or at the library. You carefully check the lounge and John is not there. If you assume that his roommate is truthful, what do you conclude? What type of reasoning are you using?

3. Aspirin has relieved every headache you have had. You now have a headache. What would you normally conclude? What type of reasoning are you using?

4. You must attain a 3.5 over-all average out of a possible 4.0 to graduate with honors. You have completed all requirements for graduation with a 3.15 over-all average. What do you conclude? What type of reasoning are you using?

5. In your everyday life, which do you use more often, inductive or deductive reasoning? Give some examples of conclusions you have reached inductively. Have these conclusions held up under the weight of subsequent events? Can you find any examples of deductive reasoning in your everyday life? If so, can you identify the axioms upon which you based one of these deductive arguments?

6. We have pointed out that axioms are statements about undefined concepts. By examining Euclid's first five axioms, you should be able to identify some undefined concepts. (For example, Axiom 1 involves the undefined concepts of *point* and *line*.) Make a list of the undefined concepts involved in these first five axioms.

7. Is the following set of axioms consistent? Explain.

> Axiom 1: $8 \cdot 2 = 10$.
> Axiom 2: $5 \cdot 2 = 6$.
> Axiom 3: $10 \cdot 6 = 4$.
> Axiom 4: $(8 \cdot 2) \cdot (5 \cdot 2) = 44$.
> Axiom 5: Equals multiplied by equals are equal.

*8. A set of axioms is called **dependent** if one of the axioms can be deduced from the other axioms of the set. Explain why the following sets of axioms are dependent.

    (a) The sum of two even numbers is even.
        The sum of two odd numbers is even.
        $2 + 4$ is even.

    (b) A square is a rectangle.
        A rectangle is a quadrilateral.
        A square is a quadrilateral.

(c) $8 \cdot 2 = 10.$
  $5 \cdot 2 = 6.$
  $(8 \cdot 2) \cdot (5 \cdot 2) = 4.$
  $10 \cdot 6 = 4.$
  Equals multiplied by equals are equal.

9. A dictionary must give a definition for every word. One dictionary contains these entries:

> Memorial Day: See Decoration Day
> Decoration Day: See Memorial Day

Does this dictionary provide a definition for Memorial Day?

10. The scientific name for a panda is *Ailuropoda melanoleuca*. The second part of this name means "black and white." The first part means "the animal with the pandalike front foot." So, according to its name, a panda is a black and white animal with a foot like a panda. Discuss the circularity of this definition.

1.2 Axiomatic Mathematics, II†

Euclid chose his axioms with care, and to him each of them represented what he might have called an "absolute truth" or a "self-evident truth" or an "undeniable truth." This view was held for 2000 years, and it was not until the early nineteenth century that the possibility of absolute and undeniable truth was called into question. We shall trace through this story in this section.

Euclid believed that he had chosen as axioms statements that were not only very primitive in what they said but also expressed ideas that must be undeniable. For him there could be no question but that each of these axioms was true of the material world; thus, there was no reason why they should not be accepted as axioms. But the fifth postulate, which is called the **Unique Parallel Postulate,** did seem to be somehow "different" from the other axioms. The Unique Parallel Postulate seemed more sophisticated than the other axioms; it seemed to express a higher order of truth. In fact, this fifth postulate looked more like a theorem than an axiom. The question was, could the fifth postulate be proved on the basis of the other postulates? If it could, then the fifth postulate would be dependent upon the other postulates. But if it could not be proved on the basis of the other postulates, then it would be independent of them. So the question was, is the fifth postulate independent of or dependent upon the other axioms? As far as we can tell this question was attacked continuously from Euclid's time to about the year 1825.

One of the men who tried to decide this question was Janos Bolyai (Hungarian, 1802–1860), a professional soldier and the son of a mathematician. Bolyai was working hard on the question when his father wrote to him, "For God's sake, I beseech you, give it up. Fear it no less than sensual passion because it, too, may take all your time, and deprive you of your health, peace of mind, and happiness in life." Clearly, working on the question of dependence of the Unique Parallel Postulate was losing its fascination for some mathematicians after 2000 years. What was needed was a new insight into the problem.

About this same time Karl Friedrich Gauss (German, 1777–1855), the greatest mathematician of his time and one of the greatest mathematicians of all time, had been working on the problem. A professor from the University of Kazan, Nikolai Ivanovitch Lobachevski (Russian, 1793–1856) was working on it also. These three men had never met (although they were known to each other), but they were all to come to just about the same conclusion at just about the same time. Let us continue the story with Lobachevski.

Lobachevski became convinced that the Unique Parallel Postulate was, in fact, independent of the other axioms. To test out this conjecture, he replaced the Unique Parallel Postulate with the following contradictory postulate.

Multiple Parallel Postulate.
Given a line ℓ and a point P not belonging to ℓ, there can be drawn through P infinitely many lines that are parallel to ℓ. (Figure 1.2.[2])

Lobachevski's reasoning was this: If the Unique Parallel Postulate were in fact independent of Euclid's other axioms, then replacing it with this new postulate should *not* produce an inconsistent geometry. On the other hand, if the Unique Parallel Postulate were dependent upon the other axioms, then replacing it with this new axiom would introduce a contradiction into the resulting deductive system. That is, the statement "Through P there is only one parallel" would be provable from the other axioms, whereas the statement "There are infinitely many parallels through P" would be an axiom; these two statements would contradict

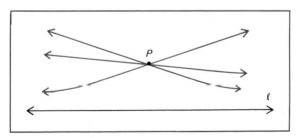

Figure 1.2

[2]This figure only hints at the true situation. Because of the smallness of the figure it has been necessary to make the infinitely many parallels appear to bend, but in actuality they do not bend at all. In order to understand the figure, it is essential that you tear yourself away from the traditional understanding that there is only one such parallel. This is difficult to do, for the assumption that there is only one such parallel line is bred into us from birth.

each other. Hence the resulting deductive system would be inconsistent. So, Lobachevski made the replacement, introducing his Multiple Parallel Postulate, and then went on to prove that the resulting deductive system was *not* inconsistent. This meant that the Unique Parallel Postulate was independent of the other axioms. Finally, after about 2000 years the independence of Euclid's fifth postulate had been established.

The deductive system that results from the assumption of Euclid's axioms (except for his Unique Parallel Postulate), together with the Multiple Parallel Postulate, is called **Lobachevskian geometry.** (As a matter of fact, both Gauss and Bolyai had come to the same conclusions as had Lobachevski, but the main credit for the invention of this new non-Euclidean geometry is usually given to Lobachevski.) But even to its creator Lobachevskian geometry appeared so unnatural that he referred to it as "imaginary geometry."

Thus the question of independence of the Unique Parallel Postulate had been solved and in the process a new kind of geometry had been invented. Here things rested for a number of years until the German mathematician Georg Riemann (1826–1866) invented another new non-Euclidean geometry, which is called **Riemannian geometry.** Riemannian geometry is based upon the assumption of Euclid's axioms (except for the Unique Parallel Postulate) plus the following axiom.

No Parallels Postulate.
Given a line ℓ and a point P not belonging to ℓ there cannot be drawn any lines through P that are parallel to ℓ. (Figure 1.3.[3])

The geometry based upon this set of axioms is also internally consistent.

There were now three distinct geometries, Euclid's (unique parallels), Lobachevski's (infinitely many parallels), and Riemann's (no parallels). From a purely axiomatic point of view each of these three geometries is as good as the other two, for all three geometries are internally consistent in the sense that none of them is capable of giving rise to a contradiction. The natural question now is, which of these geometries is the "correct" one in terms of being a "correct" abstraction of our physical environment? We cannot answer this question. The fact is that if we consider only "local" problems—that is, problems dealing with

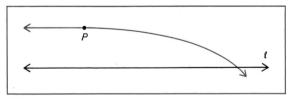

Figure 1.3

[3]The lines do not really bend as shown here. Because of the smallness of the illustration it is necessary to make them bend in order to have the point of intersection be a point on the page.

distances similar to those here on earth, say, problems in which distances do not exceed a few million miles—then these geometries give so nearly the same results that we cannot distinguish between them using our best measuring devices. For example, in Euclidean geometry the sum of the angles of a triangle is exactly 180°, in Lobachevskian geometry this angle sum is always less than 180°, while in Riemannian geometry the angle sum is always greater than 180°. Theoretically, to determine the "correct" geometry all we have to do is construct a triangle and measure the sum of its angles; then we shall know which geometry is the "real" geometry. But in order to be able to make this determination with the measuring instruments available to us, we would have to work with a triangle with sides measured not in miles but in light-years. It is doubtful that we shall ever be able to make such a determination and find out which of these geometries is "correct."

But, as Riemann realized, the invention of these non-Euclidean geometries has deeper significance than the question of which one represents our physical environment. The fact that each of these geometries is internally consistent has forced mathematicians to renounce the notion of "absolute truth" or "undeniable truth" or "self-evident truth" in mathematics. For if, as Euclid thought, his axioms were undeniable truths, this would mean that the non-Euclidean geometries are based axiomatically upon statements contrary to fact. It would then follow that these geometries would have to be internally inconsistent. But it has been proved that they are not internally inconsistent. So we must reject the idea that there are any statements about geometry that can be called absolute truths.

We now understand that truth is not an attribute that a statement has or does not have in any absolute sense. A statement that is true in one deductive system may be false in another. (For example, the statement "The sum of the angles of a triangle is 180°" is a true statement in Euclidean geometry but is false in the non-Euclidean geometries.) Thus, in order to say that a particular statement is true, we must know that the statement follows from the axioms of that system in a logically valid way. Thus by "truth" we really mean "validity." That is, a statement is true (in a given deductive system), provided that the statement can be derived in a logically valid way from the axioms of that system.

As long as mathematicians could regard the axiomatic foundations of mathematics as being undeniable truths, there was no reason to be excessively critical of those foundations. But with the destruction of the notion of absolute truth, it became clear that mathematicians would have to spend at least a part of their time examining and reexamining the axiomatic foundations of mathematics.

1.3 Introduction to Symbolic Logic

Logic was first studied in a systematic way by the ancient Greeks, who used it to provide a logical framework in which to do mathematics. But following the period of Greek mathematics, the development of logic languished for nearly 2000 years. The next great surge took place in the middle of the nineteenth century with the invention of symbolic logic. We shall discuss elementary symbolic logic in this and the next three sections.

The logic studied by the Greeks is called **Aristotelian logic** after Aristotle (about the fourth century B.C.), who was its most famous student. Aristotle produced a list of 14 syllogisms that he felt summed up the major ideas of logic. For example, one of these syllogisms is

(1) All heroes are men.
(2) All men are mortal.

Hence,

(3) All heroes are mortal.

The other syllogisms dealt with statements of the form "Some men are mortal," "No men are mortal," and so on. These syllogisms were examples of all the known ways to draw a conclusion (when such a conclusion was possible) from two given statements.

The 14 syllogisms of Aristotle, plus another five that were added by medieval logicians, represented logic for about 2000 years. To study logic essentially meant to study these 19 syllogisms. Then, in 1848, the English mathematician George Boole (1815–1864) published a little book in which he used symbols to facilitate the study of logic in much the same way that symbols are used in algebra. Moreover, Boole organized his study of what we today call symbolic logic along the lines of a deductive system. He selected undefined concepts and axioms and used these to build up the system of symbolic logic.

Just as algebra is the study of the ways that numbers can be compared and operated upon, **symbolic logic** is the study of ways that statements can be compared and operated upon. We shall leave the concept of "statement" undefined except to say that a statement is either true or false, but cannot be both true and false at the same time. We may accept this as an axiom concerning the undefined concept of "statement."

Example 1.   "Help!" is not a statement in the sense of symbolic logic because it is neither true nor false. The statement "Two plus two equals four" is a true statement and thus is one of the statements with which symbolic logic deals. So, too, is the false statement "Two plus two equals six."

Expressed as simply as possible, symbolic logic is the study of how statements can be constructed from given statements and how the truth value (that is, the truth or falsity) of the newly constructed statements can be determined from the truth values of the statements from which they were constructed. For example, suppose we are given two statements that we shall label p and q. (Note the use of symbols to denote these statements. This is typical of symbolic logic.) We can use these statements to form the compound statement "p and q." We symbolize this statement by $p \wedge q$. (Note the use of a symbol $\wedge$ to represent the word "and"; "$\wedge$" is used to denote "and" in symbolic logic in the same way that "$+$" is used to denote "added to" in algebra.) Now that we have constructed our new statement, $p \wedge q$, how can we determine its truth value in terms of the truth values of the given statements p and q? By common agreement the compound statement $p \wedge q$ is true if both p and q are themselves true and is false in all other cases. This information is contained in Table 1.1.

Example 2. The sentence "It is Wednesday and it's still snowing" is a sentence that is composed of two other sentences by means of the conjunction "and." The basic sentences are "It is Wednesday" and "It's still snowing." If we denote the first of these sentences by p and the second by q, then the original compound sentence is represented symbolically by $p \wedge q$. The sentence $p \wedge q$ is true exactly when both p and q are individually true. A sentence $p \wedge q$ would be false if p represented the statement "$2 + 3 = 7$," no matter what sentence q represented. See Figure 1.4.

The word "and" is called a **logical connective,** and the statement that results from combining two statements using this word is called a **conjunction.** A logical connective is a way of using old statements to produce new statements and corresponds to the operations of addition, multiplication, and so on, in arithmetic. There are three other basic connectives, two of which we shall discuss now. The third will be discussed in the next section.

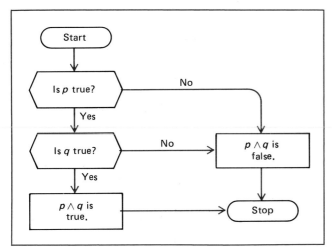

Figure 1.4 *Flow chart for determining the truth or falsity of the conjunction $p \wedge q$.*

Table 1.1

p	q	p ∧ q
T	T	T
T	F	F
F	T	F
F	F	F

If p and q are statements, we can form the compound statement "p or q," which is called the **disjunction** of p and q. The symbol representing disjunction is "$\vee$," and so we write $p \vee q$ for "p or q." We agree that the statement $p \vee q$ will be true if one or both of p and q is true but will be false when both p and q are false. See Table 1.2.

Example 3.   Let p represent the statement "I shall study for my exam" and q represent the statement "I shall go to the movies." Then the statement "I shall study for my exam or I shall go to the movies" is denoted symbolically by $p \vee q$. Therefore, $p \vee q$ will be false only if I neither study for my exam nor go to the movies.

The conjunction and disjunction connectives operate upon pairs of statements, but the **negation connective** operates upon only one statement. Given a statement p, we may form the statement "Not p." This statement is symbolized by $\sim p$ and is called the **negation** of p. The statements p and $\sim p$ have opposite truth values as shown in Table 1.3.

Example 4.   If p represents the statement "John attended class yesterday" and q represents the statement "$2 + 3 = 8$," then the negation of p is the statement "John did not go to class yesterday" and the negation of q is "$2 + 3 \neq 8$." The second of these examples illustrates that the negation of a statement may be true. Whether or not the negation of a statement is true or false depends upon whether the original statement was false or true. The statement "Not p" should not be regarded as false. It will only be false when p itself is a true statement.

Table 1.2

p	q	p ∨ q
T	T	T
T	F	T
F	T	T
F	F	F

Table 1.3

p	~p
T	F
F	T

Table 1.4

p	q	~p	~p ∨ q
T	T	F	T
T	F	F	F
F	T	T	T
F	F	T	T

These three logical connectives (together with a fourth discussed in Section 1.4) can be used in combination to construct more complex statements. For example, using the conjunction and negation connectives, we can construct the compound statement $\sim p \lor q$ from statements p and q. The truth value of such a compound statement can be most easily determined by constructing tables like those shown earlier. The table for this statement is shown in Table 1.4.

Example 5. Find statements p and q such that the statement $\sim p \lor q$ is false.

Solution: From Table 1.4 we see that the only way $\sim p \lor q$ can be false is for p to be true and q to be false. So let p be the statement "The moon is not made of Camembert" and q be the statement "Three is less than two." Then $\sim p \lor q$ is the statement "Either the moon is made of Camembert or three is less than two."

The tables that we have been using are called **truth tables.** Their principal value is that they provide a routine and generally easy way to determine the truth values of compound statements.

Example 6. When is the statement $\sim p \lor (p \land \sim q)$ true?

Solution: The truth table for this statement is shown in Table 1.5. From this table we can see that the given statement is true in all cases except when p is true and q is true.

Table 1.5

p	q	~p	~q	p ∧ ~q	~p ∨ (p ∧ ~q)
T	T	F	F	F	F
T	F	F	T	T	T
F	T	T	F	F	T
F	F	T	T	F	T

Exercises 1.3

1. Let p represent the statement "Jim is happy" and q represent the statement "Mary is sad." Consider "not happy" as "sad" and write each of the following statements symbolically:
   (a) Jim is happy or Mary is sad.
   (b) Jim is sad and Mary is sad.
   (c) Either Mary or Jim is happy.
   (d) It is not true that both Mary and Jim are sad.
   (e) Neither Jim nor Mary is happy.
2. Assume Jim and Mary are both happy. Which statements in Exercise 1 are true?
3. Let p represent the statement "$2 = 3 + 7$" and q represent the statement "Henry VIII was a king of England."
   (a) Form the statement $(p \lor q)$. Is this statement true or false?
   (b) Form the statement $\sim (q \lor p)$. What is the truth value of this statement?
4. Let p represent the statement "I like school" and q represent the statement "I like to study." Represent each of the following statements in words.
   (a) $\sim p \land \sim q$. (b) $p \land \sim q$. (c) $\sim p \lor \sim q$.
   (d) $\sim p \lor q$. (e) $\sim (p \lor q)$.
5. Construct truth tables for each symbolic statement in Exercise 4.
6. By using the truth tables constructed in Exercise 5, determine which of the statements are true if:
   (a) I like school but I do not like to study.
   (b) I neither like school nor like to study.
7. Construct a truth table for the compound statement $p \land (\sim p \lor q)$.
8. Construct a truth table for the statement $(p \land \sim q) \lor (p \land q)$. Find specific examples of statements p and q for which this statement is true.
9. Make a flow chart for determining the truth or falsity of a negation.
10. Make a flow chart for determining the truth or falsity of a disjunction.

1.4 Conditionals

The fourth logical connective is used more often than the other three and is, unfortunately, also more complicated. We shall study it in this section.

   If we are given statements p and q, we can combine them into the statement "If p, then q," by which we mean "If p is true, then q is true." This statement is called a **conditional** and is symbolized by $p \rightarrow q$. The statement p is called the **hypothesis** of the conditional $p \rightarrow q$, and q is called its **conclusion**.

Example 1.   From the given statements "$2 + 2 = 5$" and "3 is greater than 2," we can form the two conditions:

If $2 + 2 = 5$, then 3 is greater than 2.

and

If 3 is greater than 2, then $2 + 2 = 5$.

Although in everyday usage the hypothesis and conclusion of a conditional generally have some material relationship to each other, this need not be the case in symbolic logic. Any two statements can be made into a conditional, and that conditional is either true or false. For example, the conditional "If $2 + 2 = 789$, then Grover Cleveland pitched for the White Sox" is a conditional that we would probably regard as being nonsense in everyday language; however, this conditional is completely meaningful in symbolic logic. In fact, it is even true! For the axiom that establishes the truth value of a conditional in terms of the truth values of its hypothesis and conclusion is the following: The conditional $p \rightarrow q$ is false if p is true and q is false but is true in all other cases.

The full story is told in Table 1.6. If this situation appears strange, it is only because in everyday language we do not generally allow ourselves to get involved with conditionals whose hypotheses or conclusions we know to be false. Generally speaking, the average person is familiar only with conditionals whose hypotheses and conclusions he believes to be true; this is not because that is the only kind of conditional there is but because that is the only kind he ever uses. See Figure 1.5.

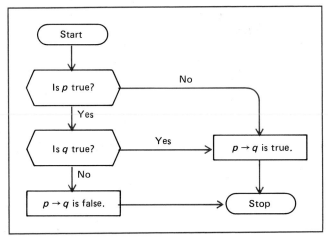

Figure 1.5 *Flow chart for determining the truth or falsity of a conditional $p \rightarrow q$.*

Table 1.6

p	q	p → q
T	T	T
T	F	F
F	T	T
F	F	T

Example 2. Which of these conditionals are false?

1. If $3 = 5$, then 2 is less than 0.
2. If 2 is less than 0, then $3 = 7 + 4$.
3. If 6 is less than 7, then gold is heavy.
4. If $2 + 4 = 7$, then $2 + 2 = 5$.
5. If $2 + 4 = 6$, then $2 + 2 = 5$.

Solution: Only the last of these conditionals is false. The only way a conditional can be false is for its hypothesis to be true while its conclusion is false; only (5) has a true hypothesis and a false conclusion.

Perhaps the most common error made regarding conditionals is thinking that because you have proved the conditional $p \to q$ is true, you have proved that the statement q is true. It is very important to understand clearly that saying $p \to q$ is true says nothing definitive about the truth values of p and q, except that it is not true that both p is true and q is false.

Given a conditional $p \to q$, we may interchange the hypothesis and conclusion to obtain a new conditional, $q \to p$, which is called the **converse** of the original conditional. From the truth table shown in Table 1.7, we can see that the truth values of a conditional and its converse are not identical. There are times when a conditional is true but its converse is false. This will be the case whenever the hypothesis of the conditional is false and its conclusion is true. Thus the true statement "If $2 + 2 = 5$, then $3 = 3$" has a false converse: "If $3 = 3$, then $2 + 2 = 5$."

Example 3. Consider the statement "If I study, then I shall pass the course." Let $p \to q$ represent the statement symbolically. If p is true and q is false (that

Table 1.7

p	q	p → q	q → p
T	T	T	T
T	F	F	T
F	T	T	F
F	F	T	T

is, I do study but do not pass the course), then the conditional is false. The converse of the statement would be "If I pass the course, then I have studied" (represented symbolically by $q \rightarrow p$). Under the assumption that q is false and p true, the converse is a true statement. If I neither study nor pass the course (that is, both p and q are false), then both the original conditional and its converse are true. What about the truth values of the conditional and its converse if I do not study but do pass the course?

Of particular importance are statements p and q such that both $p \rightarrow q$ and $q \rightarrow p$ are true. When this is the case, we write $p \longleftrightarrow q$ or "p if and only if q." In this case the statements p and q are said to be **logically equivalent.**

Example 4. The statements "$2 + 2 = 5$" and "$3 = 4$" are logically equivalent because each of the conditionals

$$(2 + 2 = 5) \rightarrow (3 = 4)$$

and

$$(3 = 4) \rightarrow (2 + 2 = 5)$$

is true. Thus any two false statements are logically equivalent. Also, any two true statements are logically equivalent because if p and q are both true, then each of the conditionals $p \rightarrow q$ and $q \rightarrow p$ is true.

To test whether two statements are logically equivalent, all we need to do is to construct truth tables for the statements and compare their truth values. If these truth values are identical, then the statements are logically equivalent.

Example 5. Are the statements $p \vee q$ and $q \vee p$ logically equivalent?

Solution: They are. We acknowledge this equivalence whenever we realize that such statements as, "I'm going to buy black shoes or I'm going to buy brown shoes" and "I'm going to buy brown shoes or I'm going to buy black shoes" have the same truth values. We can prove this by means of Table 1.8.

Example 6. Prove that $\sim(p \wedge q)$ and $\sim p \vee \sim q$ are logically equivalent.

Table 1.8

p	q	$p \vee q$	$q \vee p$
T	T	T	T
T	F	T	T
F	T	T	T
F	F	F	F

Table 1.9

p	*q*	*p* ∧ *q*	~(*p* ∧ *q*)	~*p*	~*q*	~*p* ∨ ~*q*
T	T	T	F	F	F	F
T	F	F	T	F	T	T
F	T	F	T	T	F	T
F	F	F	T	T	T	T

Solution: We use the truth table, Table 1.9. Because the entries in the columns under the statements ~(*p* ∧ *q*) and ~*p* ∨ ~*q* are identical, these statements are equivalent.

The equivalence ~(*p* ∧ *q*) ⟷ ~*p* ∨ ~*q* is known as the first **De Morgan law.** The second De Morgan law is the equivalence ~(*p* ∨ *q*) ⟷ ~*p* ∧ ~*q* (which we leave for you to prove by using truth tables). These equivalences are named after the English mathematician Augustus De Morgan (1806–1871), who along with Boole shares the major credit for the invention of symbolic logic. The first De Morgan law states that the denial of conjunction is a disjunction, and the second law states that the denial of a disjunction is a conjunction.

Example 7. Assume Ethelbert was lying when he promised to love *and* honor his new wife. Then De Morgan's first law tells us that *either* he didn't love her *or* he didn't honor her.

Example 8. Assume Ethelred lied when he promised his bride that he would *either* take her to New York *or* to New Orleans for their honeymoon. Then De Morgan's second law tells us that he didn't take her to New York *and* he didn't take her to New Orleans.

Exercises 1.4

1. Make a truth table that proves that the statements *p* → *q* and ~*p* ∨ *q* are logically equivalent. Use this fact to explain why the statements "If it rains today, then the ground will get wet" and "Either it will not rain today or the ground will get wet" express the same idea.
2. If the conditional *p* → *q* and its converse are both true, then how do the truth values of *p* and *q* compare? If the conditional and its converse have opposite truth values, how do the truth values of *p* and *q* compare? (*Hint:* Use Table 1.7.)
3. By using truth tables, decide whether the following are logically equivalent:
   (a) *p* ∧ *q* and ~*q* → *p*.
   (b) ~(*p* ∧ *q*) and *p* → *q*.

(c) $\sim p \to (p \lor q)$ and $p \lor \sim q$.

(d) $\sim p \lor \sim q$ and $p \to (\sim q \land p)$.

*4. Use a truth table to prove that the statements $p \to q$ and $\sim q \to \sim p$ are logically equivalent. The conditional $\sim q \to \sim p$ is called the **contrapositive** of the conditional $p \to q$.

*5. Because (by Exercise 4) the contrapositive $\sim q \to \sim p$ of a conditional $p \to q$ is logically equivalent to the conditional itself, we can replace a conditional by its contrapositive whenever we feel that it would be convenient to do so. Replace each of these statements by its contrapositive.

(a) If $x + 3 = 7$, then $x = 4$.

(b) if $x + 3 \neq 7$, then $x \neq 4$.

(c) If it rains, then the ground will get wet.

(d) If the diagonals of a rectangle do not intersect at right angles, then that rectangle is not a square.

(e) If a number is not a multiple of two, then it is not an even number.

*6. Make a truth table for the statement $(p \to q) \lor \sim q$. Note that the entries in the column under this statement are all T's. This means that this statement is true no matter what the truth values of the component statements p and q might be. Such statements are called **tautologies**. Make truth tables for the following statements and prove that each is a tautology.

(a) $p \lor (p \to q)$.   (b) $(p \to q) \lor (p \to \sim q)$.

(c) $(p \land q) \lor (p \to \sim q)$.

*7. Which of these statements are tautologies?

(a) $p \to (p \land q)$.   (b) $(p \to q) \lor (q \to p)$.

(c) $(p \land \sim q) \lor (q \land \sim p)$.

*8. The **inverse** of a conditional is obtained by negating both hypothesis and conclusion. For example, the inverse of $p \to q$ is $\sim p \to \sim q$. Find an example of a conditional whose inverse is true.

*9. Write the converse, contrapositive, and inverse of each of the following conditionals:

(a) If I go shopping, I shall spend money.

(b) If I arise early, I shall be healthy, wealthy, and wise.

(c) If I take a stitch in time, I shall save nine.

(d) If I eat an apple a day, the doctor will stay away.

(d) A rolling stone gathers no moss. (*Hint:* First express in the form of a conditional.)

10. Make a truth table that proves the second De Morgan law.

11. Use the De Morgan laws to replace each of these false statements with a true statement.

(a) $2 < 3$ and $2 > 3$.

(b) The next president will be both a Democrat and a Republican.

(c) The solution of the equation $x + 3 = 5$ is either 6 or 7.

(d) Either Henry Clay was president or he was born in Kentucky.

1.5 Arguments

The mathematical proof of a theorem is a demonstration using logical arguments that the truth of that theorem follows from the established truth of previously proved theorems and the assumed truth of the axioms. In this section we shall discuss a few of the most important and useful of these logical arguments.

Logical arguments are based upon the assumption of certain statements called "hypotheses." One assumes the hypotheses and then argues from them using laws of logic so as to obtain the statement that is the conclusion of the argument. For example, Aristotle's syllogism,

(1) All heroes are men.
(2) All men are mortal.

Therefore,

(3) All heroes are mortal.

argues from the assumption of statements (1) and (2) as hypotheses and using a law of logic arrives at the conclusion (3). We must emphasize, however, that this argument does not establish the truth of any one of the statements (1), (2), or (3). Indeed, in some arguments all the hypotheses as well as the conclusion will be false statements. What the argument does do is to prove that the statement

$$[(1) \wedge (2)] \rightarrow (3)$$

is true.

But what law of logic has permitted us to draw conclusion (3) from the assumption of hypotheses (1) and (2)? By using truth tables, we can prove that the statement

$$[(p \rightarrow q) \wedge (q \rightarrow r)] \rightarrow (p \rightarrow r)$$

is true no matter which statements p, q, and r represent. This statement is called the **law of syllogism.** If we rewrite Aristotle's argument, we can see how this law of syllogism was used to establish its validity:

(1) If a thing is a hero, then that thing is a man.
(2) If a thing is a man, then that thing is mortal.

Therefore,

(3) If a thing is a hero, then that thing is mortal.

Table 1.10

p	q	r	$p \to q$	$q \to r$	$(p \to q) \wedge (q \to r)$	$p \to r$	$[(p \to q) \wedge (q \to r)] \to (p \to r)$
T	T	T	T	T	T	T	T
T	T	F	T	F	F	F	T
T	F	T	F	T	F	T	T
T	F	F	F	T	F	F	T
F	T	T	T	T	T	T	T
F	T	F	T	F	F	T	T
F	F	T	T	T	T	T	T
F	F	F	T	T	T	T	T

We prove the law of syllogism by constructing a truth table (refer to Table 1.10). Observe that the entries in the rightmost column are all T's. This means that the law of syllogism is true no matter what the truth values of its component statements p, q, and r might be.

The law of syllogism is a way to "glue" two conditionals together when the conclusion of one is the same as the hypothesis of the other. By repeated applications of this law a string of conditionals can be combined.

Example 1.   Assume the following statements as hypotheses:

(1) If a man smokes, then he will get cancer.
(2) If a man gets cancer, then he must go to the hospital.
(3) If a man goes to the hospital, then he must pay large bills.
(4) If a man must pay large bills, then he must work longer hours.

Hence,

(5) If a man smokes, then he must work longer hours.

When we make this argument, we do not assert that any of the hypotheses are true or that the conclusion is true (although in everyday situations we probably wouldn't bother with the argument unless these statements were believed to be true), but we do assert that the conditional

$$[(1) \wedge (2) \wedge (3) \wedge (4)] \to (5)$$

is true. The symbolic form of this argument is

(1) $p \to q$.
(2) $q \to r$.
(3) $r \to s$.
(4) $s \to t$.

Therefore,

(5) $p \rightarrow t$.

The law of syllogism is the basis for many logical arguments used in mathematics. Another law that is the basis for many arguments is the **law of detachment:**

$$[p \wedge (p \rightarrow q)] \rightarrow q.$$

This law asserts that if we assume the truth of the conditional $p \rightarrow q$ and if we also assume the truth of p, then we must be prepared to accept the truth of q as well.

Example 2. If we accept the truth of the statement "If John gets married, then he is going to buy a house" and "John got married last Saturday," then we are forced to accept the truth of the statement "John is buying a house."

Example 3. If we add a fifth hypothesis to the argument in Example 1, "I smoke," then we may draw the conclusion that "I must work longer hours."

The symbolic form of the arguments in both Example 2 and Example 3 is

(1) $p \rightarrow q$.
(2) p.

Therefore,

(3) q.

Table 1.11 shows that the law of detachment is true no matter what the truth values of the component statements p and q might be. This proves that the law of detachment is a valid law of logic.

There are many laws of logic. We have already discussed the law of syllogism and the law of detachment. Here is a short and incomplete list of some other laws of logic. These laws are called **tautologies.** (See Exercise 6 in Section 1.4.)

Table 1.11

p	q	$p \rightarrow q$	$p \wedge (p \rightarrow q)$	$[p \wedge (p \rightarrow q)] \rightarrow q$
T	T	T	T	T
T	F	F	F	T
F	T	T	F	T
F	F	T	F	T

1. $p \lor \sim p$ (law of the excluded middle). This law expresses something we have already assumed about statements—that they are either true or false. It is called the law of the excluded middle because it states there is no "middle ground" for a statement; it must be either true or false, and there is no other possibility.

2. $\sim(\sim p) \longleftrightarrow p$ (law of double negation). For example, if p represents the statement "I am going home," then $\sim(\sim p)$ represents the statement "It is false that I am not going home," which, as we all recognize (eventually), is just another way of saying p.

3. $\sim(p \land \sim p)$ (law of contradiction). Like Tautology 1, this tautology expresses something we have assumed about statements. It says that it is impossible for a statement to be both true and false at the same time.

4. $\sim(p \land q) \longleftrightarrow \sim p \lor \sim q$ and $\sim(p \lor q) \longleftrightarrow \sim p \land \sim q$. These are the two De Morgan laws that we discussed in the last section.

5. $(p \land q) \to p$. This tautology simply states that if the statement $p \land q$ is true, then p is true. We already know that this is so; and, in fact, we know more—we know that q is true as well.

6. $q \to (p \to q)$. This tautology states that if the conclusion of a conditional is true, then the conditional itself must be true. If you will consult the truth table for conditionals, you will see that every time the conclusion of a conditional is true the conditional itself is true. In other words, any statement at all implies a true statement. [Don't confuse this with the fact that a false statement implies any statement at all: $\sim p \to (p \to q)$.]

7. $(p \to q) \longleftrightarrow (\sim q \to \sim p)$. This tautology states that a conditional is logically equivalent to its contrapositive. (See Exercises 4 and 5 in Section 1.4.)

There are more tautologies in the exercises that follow.

Exercises 1.5

1. Determine whether or not each argument is valid. If valid tell which tautology you used.
   (a) Given: The statement "John went to the basketball game or he went to the doctor's" is false.
       Conclusion: John did not go to the doctor's.
   (b) Given: I like peanut butter.
       Conclusion: If cows have three legs, then I like peanut butter.
   (c) Given: If I eat googols, then I will be strong. I am not strong.
       Conclusion: I do not eat googols.
2. Use the laws of syllogism and detachment either together or singly to draw conclusions from the sets of hypotheses that follow.
   (a) $(2 < 3) \to (3 < 4)$.
       $2 < 3$.
       $(3 < 4) \to (4 < 5)$.

(b) If horses give milk, then people ride on cows.
     If people ride on cows, then they will get sore.
     If people get sore, then they get mad.

(c) $p \rightarrow q$.
    $q \rightarrow r$.
    p.
    $r \rightarrow s$.

(d) $(p \wedge q) \rightarrow (p \vee q)$.
    $(r \rightarrow s) \rightarrow (p \wedge q)$.
    $r \rightarrow s$.

3. From the hypotheses "If $x + 3 = 5$, then $x = 2$" and "If $x = 2$, then $x + 1 = 3$," can you draw the conclusion "$x + 1 = 3$"? What conclusion can you draw? What additional hypothesis do you need in order to be able to draw the conclusion "$x + 1 = 3$"?

*4. The conditional

$$[(p \rightarrow q) \wedge (\sim q)] \rightarrow \sim p$$

is called the law of **modus tollendo tollens.** The law is also the basis for many arguments. Use it to draw a conclusion from the given hypotheses.

(a) $(2 + 2 = 5) \rightarrow (2 = 3)$.
    $2 \neq 3$.

(b) If it is morning, the sun is on the eastern horizon.
    The sun is on the western horizon.

(c) $(3 + 5 < 9) \rightarrow$ (horses fly).
    Horses do not fly.

(d) $(2 + 2 > 5) \rightarrow (3 + 5 < 9)$.
    $(3 + 5 < 9) \rightarrow (2 \neq 5)$.
    $2 = 5$.

*5. Another law of logic is **modus tollendo ponens:**

$$[(\sim p) \wedge (p \vee q)] \rightarrow q.$$

Use this law to draw a conclusion from the given hypotheses.

(a) Cows either give milk or fly.
    Cows do not give milk.

(b) $(2 + 2 < 3) \vee (3 < 4)$.
    $3 > 4$.

(c) Either Ethelred was king or Ethelbert was not king.
    Ethelred was not king.

(d) Either John works at the factory or Mary goes to school.
    Either John does not work at the factory or Harry is retired.
    Harry works at the factory.

*6. The laws of syllogism and detachment, together with two laws in Exercises 4 and 5 (modus tollendo tollens and modus tollendo ponens), are sufficient to enable you to draw a conclusion from each set of hypotheses given. Do so and explain where you use any of these laws in drawing your conclusion.

(a) If $x \neq 0$, then $x \neq y$.
   Either $x = y$ or horses fly.
   Horses do not fly.
(b) Either $3 = 4$ or cows fly.
   If cows fly, then so do horses.
   Horses do not fly.
(c) If horses live in houses, then $x = 0$.
   Either horses live in houses or John loves Mary.
   If John loves Mary, then Sue is unhappy.
   Sue is smiling and singing.
(d) If horses fly, then cows give milk.
   If cows give milk, then $2 < 3$.
   Either horses fly or John loves Mary.
   If John loves Mary, then Sue is unhappy.
   Sue is dancing and laughing.

1.6 Universal and Existential Statements

There are two kinds of statements that because of their special internal form are of particular importance in mathematics. In this section we shall discuss these two kinds of statements.

Statements whose general form is "All A's are B's" are called **universal statements.** Such statements may appear in many guises, the following being some of the ways the universal statement "All men are mortal" can be phrased:

1. Every man is mortal.
2. Each man is mortal.
3. If a thing is a man, then that thing is mortal.
4. No matter which man we consider, that man is mortal.
5. There are no men who are not mortal.

To prove a universal statement is false, all we need to do is to exhibit an example of its falsity. For example, we can prove that the statement, "Every man is under 6 feet tall" is false simply by pointing to a tall man. An example of the falsity of a universal statement is called a **counterexample,** and all that is needed to prove such a statement false is to exhibit a counterexample.

Example 1.   In order to prove the universal statement "All numbers are greater than 10" is false, all that is necessary is to cite the counterexample 2 (or any number less that 10).

Example 2. Suppose you are told "All brown cows give chocolate milk." If you could exhibit one brown cow that did not give chocolate milk, this would be a counterexample and prove the statement false.

The trouble comes when we try to prove that a universal statement is true. A common error is to exhibit a number of instances of the truth of the universal statement and then to think that, by so doing, we have proved the statement is true. Actually exhibiting such examples does not help at all to prove the statement is true. For example, we can find infinitely many numbers that are even, but this does not mean that the universal statement "All numbers are even" is true. In order to prove that a universal statement is true, we must somehow prove that no counterexamples can exist. There are a number of ways that this can be done. We shall consider some of them in our work in future chapters.

The second type of statement that we want especially to consider is the existential statement. An **existential statement** asserts the existence of something. For example, the statement "There exists a man who is mortal" is an existential statement. Like universal statements, existential statements may appear in many guises. The existential statement above can also be phrased in these ways:

1. At least one man is mortal.
2. A few men are mortal.
3. Some men are mortal.
4. One or more men are mortal.

In particular, note that "some" and "few" are used synonymously with "at least one" and "one or more"; that is, if one man is mortal, then the statement "Some men are mortal" is true. This is in partial variance with the way these words are used in everyday life.

As to proving the truth or falsity of existential statements, existential statements are easier to prove true than to prove false. All that is needed to prove that an existential statement is true is to exhibit an example of its truth. We can prove that the statement "Some men are mortal" is true by giving the name "Grover Cleveland." (The exhibition of more than one example is not necessary. One example of the truth of an existential statement is enough to prove that the statement is true.) To prove that an existential statement is false, however, we must somehow prove that no examples of its truth are possible. This is, in general, harder to do. We shall see a few different ways to do this in future chapters.

Let us close this discussion with the observation that the negation of a universal statement is an existential statement and the negation of an existential statement is a universal statement. The following examples should help to make this clear.

Example 3. If it is false that all men are mortal, then what is true?

Solution: If it is false that all men are mortal, then there must be a counterexample. This counterexample is a man who is immortal. Hence to say that it is false that all men are mortal is the same as saying that there exists a man who is immortal. So the negation of the universal statement "All men are mortal" is the existential statement "Some men are immortal."

Example 4.   If it is false that some men are mortal, then what is true?

Solution: If it is false that some men are mortal, then it is impossible to find an example of a man who is mortal. Hence all men are immortal. So the negation of the existential statement "Some men are mortal" is the universal statement "All men are immortal."

Exercises 1.6 _____

1. Rephrase each of the following statements as a statement involving the word "Every" or the phrase "There exists."
   (a) Some men work hard and a few men work very hard.
   (b) There are men who do not work hard.
   (c) At least one man works hard, but there are no men who do not sleep.
   (d) No matter which number n is, $n + 1 = 1 + n$.
   (e) There is at least one number that, when squared, is equal to itself.
2. Write out a true existential statement corresponding to each of these false universal statements.
   (a) Every number is greater than 0.
   (b) Every dog barks.
   (c) Every conditional has a true converse.
3. Write out a true universal statement corresponding to each of these false existential statements.
   (a) Some conditionals are neither true nor false.
   (b) There is a man who is totally without value.
   (c) Some true conditionals have false contrapositives.
4. Search through this book and find two universal and two existential statements. Write out the negation of each statement.
5. Exhibit a counterexample that proves that each of the universal statements in Exercise 2 is false.
6. Exhibit an example that proves each of these existential statements is true.
   (a) There exists a number that is greater than 4, is less than 10, and is even.
   (b) Some conditionals have true converses.
   (c) Some existential statements are true.
7. When we make a statement of the form "Every A is a B," are we necessarily implying that there are any A's? That is, does this universal statement assert the existence of any A's?

Mathematical Systems

The study of mathematics consists of the study of many different mathematical systems. You are more or less familiar with a mathematical system called the arithmetic of the whole numbers and with another called Euclidean geometry. These two mathematical systems differ most obviously in that the first deals with numbers and the second deals with geometric objects, such as lines and circles. If we were to center our attention too strongly upon the differences between numbers and geometric objects, we would have great difficulty in trying to see the ways that the study of numbers is similar to the study of geometric objects. In the early days of mathematics (say, prior to 1850) the particular objects being studied (numbers, geometric objects, equations, and so forth) received so much focused attention that the similarities between these various mathematical systems went unnoticed. Then, after a time, it became increasingly clear that even though the various parts of mathematics dealt with different kinds of objects, if one were to pay less attention to the particular nature of the objects themselves and concerned himself more with the way that he studied the objects, then he would see patterns to these various studies and that, to a greater or lesser extent, these patterns are similar.

Our objective in this chapter is to inquire into the general structure of mathematical systems. The particular nature of the objects of the systems (whether they be numbers, geometric objects, equations, or whatever) will be of secondary importance to us here.

2.1 Mathematical Systems: A General Introduction

To introduce the notion of a mathematical system, we shall use as our example the mathematical system with which you are probably most familiar, the mathematical system of whole numbers. The whole numbers are the numbers 0, 1, 2, 3, 4, 5, and so on. When we think of doing arithmetic with whole numbers, we naturally first think of these numbers themselves. But what other ideas are involved besides the idea of a whole number? Let us list a few of these other ideas.

1. The idea of one whole number being equal to another.
2. The idea of one whole number being less than another.
3. The idea of one whole number being equal to the square of another.
4. The idea of adding two whole numbers to obtain a third.
5. The idea of multiplying two whole numbers to obtain a third.
6. The idea of dividing one whole number by another to obtain a third.

We can classify these ideas into two groups. The first three ideas have to do with relationships between whole numbers. These ideas involve comparing one whole number with another. The last three ideas are quite different. Each of these involves using given numbers to produce a new number. Thus we see that in studying the system of whole numbers we do two entirely different kinds of things. One of these things is that we compare numbers with each other. We take two numbers and ask in what ways they are related to each other. For instance, in what ways are the numbers 3 and 4 related to each other? Well, 3 is related to 4 in that it *is less than* 4; 3 is related to 4 in that 3 *is equal to one less than* 4; 3 is related to 4 in that it *is greater than the positive square root of* 4; and so on. In fact, these two numbers are related to each other in very many ways indeed.

The second thing we do with numbers is to *operate* upon them to produce new numbers. For example, how can we operate upon 3 and 4 to produce a new number? We can operate upon 3 and 4 by *adding* them to obtain 7; by *multiplying* them to obtain 12; by *subtracting* 3 from 4 to obtain 1; and so on.

The moral of this is that when we study the arithmetic of whole numbers, we do more than simply look at whole numbers. We *compare* the numbers and we *operate* upon them to produce other whole numbers. This demonstrates that the mathematical system of the whole numbers involves three essentially different ideas:

1. The set of objects of the system (that is, the whole numbers).
2. Relations by means of which these objects may be compared with one another.
3. Operations that can be used to produce new numbers from old numbers.

To further illustrate these three parts of a mathematical system, let us consider some other examples of such systems.

Example 1.   Suppose that we were to try to organize the study of straight lines in the plane into a mathematical system. We have already identified the objects of the system—the straight lines in the plane. What are some of the relations that we might want to investigate? That is, what are some of the important ways that straight lines may be compared? Probably, the relations *is parallel to* and *is perpendicular to* spring to mind. In Figure 2.1 the black lines are related to each other by the relation *is parallel to,* whereas the lines in color are related to each other by the relation *is perpendicular to.* Of course, there are many other relations, such as *is coincident with, forms a 30° angle with,* and so on.

What might be some useful and important operations on straight lines? That is, if you were given two straight lines in the plane, how might you use these lines to produce a third line? Well, if the lines intersect and form an acute angle (an angle less than 90°), then you could use the given pair of lines to produce the line that is the bisector of the acute angle. In Figure 2.2 the line in color is the bisector of the acute angle formed by the two black lines. Note, however, that if the given pair of lines had been parallel, no third line could have been produced in this way.

Example 2.   The symbolic logic we studied in Chapter 1 is organized as a mathematical system, although we did not stress this in that chapter. Among the important relations of symbolic logic we studied is the relation *is logically equivalent to.* Some of the operations of symbolic logic are conjunction, disjunction, and implication. Each of these three is a means of using two given statements to produce a third statement. Note that negation is not the same kind of operation as are the other three. Conjunction, disjunction, and implication are called *binary* operations because they act on *two* statements to produce a third. Negation acts upon only one statement to produce a second and so is called a *unary* operation.

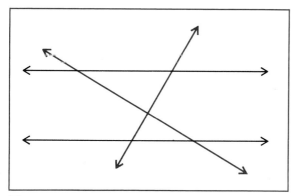

Figure 2.1

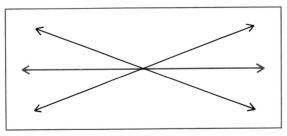

Figure 2.2

A mathematical system involves relations and operations, and in the coming sections we shall discuss these ideas more thoroughly.

Exercises 2.1

1. Which of the following describe relations on whole numbers and which describe operations? (x and y denote whole numbers.)
   (a) The idea of x being a divisor of y.
   (b) The idea of finding the least common multiple of x and y.
   (c) The idea of multiplying x by itself y times.
   (d) The idea of x being greater than six times y.
   (e) The idea of x being equal to the square root of y.

2. The numbers 5 and 25 may be compared with each other using many different relations. For example, 5 is related to 25 by the relation *is less than*. Name two other relations that relate 5 to 25. Name two relations by which 25 is related to 5.

3. The number 3 compares to 5 in the same way that 9 compares to 11. What is the relation that is being used here? What number is related to 7 by this relation? To what number is 7 related by this relation? Is 0 related to any whole number by this relation? Is any whole number related to 0 by this relation?

4. Find three different pairs of values for whole numbers x and y if x is related to y by the relation *is less than the square root of*.

5. Name two different relations that would be of interest in the study of triangles in geometry.

6. Draw two different rectangles which are related to each other by the relation *has the same area as*. Is every rectangle related to itself by this relation? If rectangle $\mathcal{R}_1$ is related to rectangle $\mathcal{R}_2$ by this relation, is $\mathcal{R}_2$ related to $\mathcal{R}_1$ also? If rectangle $\mathcal{R}_1$ is related to rectangle $\mathcal{R}_2$ and if $\mathcal{R}_2$ is related to rectangle $\mathcal{R}_3$ by this relation, is rectangle $\mathcal{R}_1$ related to rectangle $\mathcal{R}_3$?

7. Name three relations that might be of interest in a discussion of the men who have been presidents of the United States. (One such relation is *served longer than*.)

8. The number 14 is produced from the numbers 49 and 2 in the same way that

24 is produced from 36 and 4. Describe in words how this operation is performed. What is the effect of this operation upon 25 and 36? Is the effect of the operation upon 25 and 36 the same as the effect of the operation upon 36 and 25?

9. Can the "acute angle bisector" operation discussed in the text be applied to *every* pair of lines in the plane?

2.2 Relations in a Mathematical System

We have said that by a relation on the set of whole numbers we mean a way of comparing whole numbers. For example, the relation *is less than* is a relation on the set of whole numbers because this idea can be used to compare whole numbers. The whole number 3 compares with the whole number 6 using this relation (3 is less than 6), but 7 does not compare with 5 using this relation (7 is not less than 5). Another familiar relation is *equality* of whole numbers. To say that two whole numbers are equal means that they are related to each other by the relation *is equal to*. We usually define equality of whole numbers like this: If n and m are symbols representing two whole numbers, then when we write $n = m$ we mean that n and m represent the same whole number. For example, $4 - 1$ and $1 + 2$ are symbols representing the same whole number, and so we write $4 - 1 = 1 + 2$.

The notion of a relation is not restricted just to sets of mathematical objects. The idea is perfectly general and is applied to all kinds of sets of objects.

Example 1. Consider the set of all people and the relation called *is the father of*. Given two people x and y, we can ask the question: "Is x related to y in that x is the father of y?" If x is Edward I of England and y is Edward II, then the answer to this question is yes. If x is Theodore Roosevelt and y is Franklin Roosevelt, then the answer is no. Is anyone related to himself by this relation?

Consider the relation of Example 1 above. Instead of saying

Edward I is related to Edward II by the relation *is the father of,*

we would ordinarily say

Edward I is the father of Edward II,

which may be shortened even more if we simply write

Edward I (father) Edward II,

which may finally be abbreviated ultimately by

Edward I 𝔉 Edward II,

if we understand that the symbol 𝔉 means *is the father of*. We often abbreviate such statements by using a single symbol in this way. For example, instead of saying "the number named by $1 + 1$ is the same as the number named by $3 - 1$," we employ the special symbol $=$ and write simply $1 + 1 = 3 - 1$. To say that the whole number 3 is less than the whole number 4, we use the special symbol $<$ to mean "is less than" and write the abbreviation $3 < 4$. Since 4 is not less than 3, we write $4 \not< 3$. This use of the stroke to mean "is not related to by the relation $<$" is typical. The symbol $\neq$ means "is not equal to."

Example 2. There is a commonly used relation on the playing cards in a standard 52-card deck. This relation might be called the *has higher rank than* relation. If we temporarily use the symbol $>$ to mean *has higher rank than,* then we can indicate the way that the cards are related to each other by this relation as follows:

$$A > K > Q > J > 10 > 9 > 8 > 7 > 6 > 5 > 4 > 3 > 2.$$

Thus aces are the highest ranking cards, and deuces are the lowest ranking cards. The suit of a card plays no role in this relation.

Example 3. Consider the set of all triangles and the relation *is congruent to*. If we use the symbol $\cong$ to denote this relation and if $\mathcal{A}$, $\mathcal{B}$, $\mathcal{C}$, and $\mathcal{D}$ are the triangles shown in Figure 2.3, then the following statements are all true:

$$\mathcal{A} \cong \mathcal{A}, \quad \mathcal{C} \cong \mathcal{D}, \quad \mathcal{A} \not\cong \mathcal{B}, \quad \mathcal{A} \not\cong \mathcal{D}.$$

Let us return now to the relation of equality of whole numbers. Recall that to write "$n = m$" means that the symbols n and m represent the same whole number. There are three important properties of this relation that we shall use often in our study.

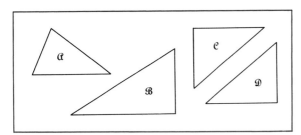

Figure 2.3

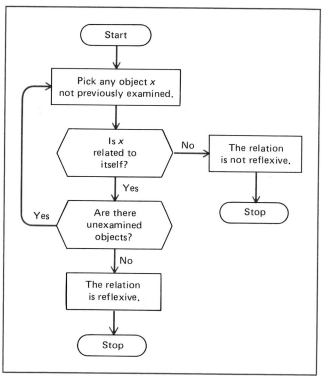

Figure 2.4 *Flow chart for determining whether a relation defined on a finite set is reflexive.*

1. *The Reflexive Property of Equality.* No matter which whole number n represents, n is equal to itself: $n = n$. (See Figure 2.4.)
2. *The Symmetric Property of Equality.* If we know that $n = m$, then we may conclude that $m = n$.
3. *The Transitive Property of Equality.* If we know that $n = m$ and that $m = p$, then we may conclude that $n = p$.

Because the relation of equality possesses these three properties, we call equality an **equivalence relation.** More generally, if a relation $\Re$ has the following properties:

1. No matter which object n represents, $n\Re n$.
2. If we know that $n\Re m$, we may conclude that $m\Re n$.
3. If we know that $n\Re m$ and $m\Re p$, we may conclude that $n\Re p$.

then we call $\Re$ an equivalence relation.

Example 4. Consider the set of all lines in the plane and the relation *is parallel to*. This relation is an equivalence relation. First, the relation is reflexive because every line is parallel to itself. Second, the relation is symmetric because, if a line l_1 is parallel to a line l_2, then the line l_2 is also parallel to the line l_1. Third, if line l_1 is parallel to line l_2 and if line l_2 is parallel to line l_3, then we may conclude that line l_1 is parallel to line l_3. Therefore, the relation is transitive.

Example 5. The relation *is perpendicular to* defined on the set of all lines in the plane is not an equivalence relation because the relation is not reflexive. That is, it is false that every line is perpendicular to itself. However, the relation is symmetric because, if line l_1 is perpendicular to line l_2, it follows that line l_2 is perpendicular to line l_1. Why is the relation not transitive?

Example 6. Which of the three properties is possessed by the relation *is less than* defined on the set of whole numbers?

Solution: Because not every whole number is less than itself, the relation is not reflexive. Because $x < y$ does not imply $y < x$, the relation is not symmetric either. But if $x < y$ and $y < z$, it does follow that $x < z$; thus, the relation is transitive.

Exercises 2.2

1. Is the relation *is the father of* an equivalence relation?
2. What properties does the relation *is the sibling of* have that neither of the relations *is the brother of* or *is the sister of* has?
3. The relation *is an ancestor of* has only one of the properties of an equivalence relation. Which one is this?
4. Is the relation *is logically equivalent to* an equivalence relation on the collection of all statements?
5. Can you think of a relation defined on people that is reflexive but is neither symmetric nor transitive?
6. What properties does the relation *is equal to or is equal to one more than* defined on the set of whole numbers possess?
7. Consider the set of all professional hockey players. Describe two relations defined on the objects of this set that are symmetric.
8. Look up the words *reflexive, symmetric,* and *transitive* in the dictionary. Can you see why these properties were named the way they were?
9. Find a relation on the set of all triangles that is neither reflexive nor symmetric but is transitive.
10. Which of the three properties does the relation *is equal to one more than or is equal to one less than* defined on the set of all whole numbers possess?
11. Explain why the flow chart in Figure 2.4 works only for a relation defined on a finite set of objects.
12. Make a flow chart for determining if a relation defined on a finite set is symmetric.

2.3 Operations in a Mathematical System

The third aspect of a mathematical system is the idea of an operation. The concept is not particularly complicated, but perhaps it would be best to begin with a familiar example. On the set of whole numbers, there is defined a particular operation called addition. What kind of a process is addition of whole numbers? First, addition works on *pairs* of whole numbers and so is called a *binary* operation. Second, this process may be applied to each and every pair of whole numbers; no matter which whole numbers x and y represent, it makes sense to talk about adding x and y (in that order—first x, then y). Third, not only can the operation be applied to every pair of whole numbers, but the result of this application on any particular pair is always one and only one whole number. For example, the result of operating upon 3 and 5 is the single number 8. So there are these three aspects of the binary operation called addition of whole numbers:

1. It is applied to exactly two whole numbers at a time.
2. It may be applied to every pair of whole numbers regardless of the order in which they are given.
3. It always results in exactly one whole number; that is, the result of the operation is unique.

The process of multiplication is also a binary operation on the whole numbers. This means that no matter which whole numbers x and y represent, the multiplication operation can be applied to x and y (in that order—first x, then y) and that the result is a single whole number. Subtraction of whole numbers, on the other hand, is not a binary operation because subtraction can be applied only to certain pairs of whole numbers. For instance, subtraction can be applied to 5 and 3 (in that order) to give the single number 2, but it cannot be applied to the numbers 3 and 5 (in that order) because $3 - 5$ is not a whole number. (The result is a number, but it is not a whole number.) Similarly, division of whole numbers is not a binary operation because it is not always possible to apply the process. For example, division cannot be applied to 5 and 0 (given in that order) because $5 \div 0$ is not a whole number. (We shall see later than $5 \div 0$ is a nonsense symbol—it has no meaning at all.)

Now, let us generalize from these remarks. By a **binary operation** on a set S of objects, we mean a process that enables us to produce a single object of the set S from any pair of objects of the set S that we might be given. Let us consider some examples of binary operations defined on familiar sets of objects in order to gain an appreciation of the general nature of this idea.

Example 1. The set of objects of this example is the standard deck of 52 playing cards. Given any two of these cards, we can produce a third card according to this rule: If x and y are playing cards, then the result of the operation is that card which has the suit of x and the rank of y. Thus the operation applied to the cards 2♠ and 5♢ is the card with the suit of 2♠ and the rank of 5♢; that is, the result is the card 5♠. The operation applied to 5♢ and 2♠ is the card 2♢. This process is a binary operation because it can be applied to any given pair of cards, and the result of the process is always a single card.

We used the special symbol + to indicate the result of the addition process upon a pair of whole numbers. Thus $3 + 5$ means the result of applying the addition process to the whole numbers 3 and 5 given in that order. If we use the special symbol * to indicate the result of acting upon a pair of cards with the operation defined in Example 1, then we could write 2♠ * 5♢ to mean the result of operating upon 2♠ and 5♢. Thus 2♠ * 5♢ = 5♠, 5♢ * 2♠ = 2♢, and K♣ * A♡ = A♣.

Example 2. On the set of all *even* whole numbers the process defined by the rule "To operate upon the even whole numbers x and y (in that order), add them" is a binary operation. This is a binary operation because any two even whole numbers can be added, and the result of this process is a single even whole number.

Example 3. Is multiplication a binary operation on the set of all odd whole numbers?

Solution: Yes. The reason is that the product of any two odd whole numbers is a single odd whole number.

Example 4. Is addition a binary operation on the set of all odd whole numbers?

Solution: No. The addition of odd whole numbers is not a binary operation because it is possible to find at least one pair of odd whole numbers whose sum is not an odd whole number. For example, the sum of 1 and 3 is not an odd whole number.

The idea of a binary operation has a very wide application in mathematics; there are binary operations defined on all sorts of objects besides numbers. In the next two sections we shall discuss some properties that may be possessed by binary operations.

Exercises 2.3

1. Which of the following are binary operations?
    (a) Addition of odd whole numbers.
    (b) Multiplication of even whole numbers.
    (c) Subtraction of even whole numbers.
    (d) Multiplication of whole numbers greater than 10.

(e) Addition of one-digit whole numbers.

(f) Addition of two-digit whole numbers.

(g) Addition of pairs of the numbers 0, 3, 6, 9, 12, 15, ...

2. The *average* of two whole numbers is their sum divided by 2. Is the averaging process as applied to whole numbers a binary operation on the set of all whole numbers?

3. Because addition is a binary operation, it is only possible to add two numbers at a time. In fact, using the ordinary addition methods by which we usually compute sums, we add only one-digit numbers. How many different applications of the addition operation to one-digit numbers are needed in order to find the sum of the numbers 234, 145, 37, and 97?

4. These questions apply to the playing card operation defined in Example 1.

(a) If $x * 3\Diamond = 3\Diamond$, then what can you say about the card x?

(b) If $2\heartsuit * x = y$, then how are the cards x and y related?

(c) If $x * y = z$, then how are the cards x and z related?

(d) Is it possible to find two different cards x and y such that $x * y = x$? Such that $x * y = y$?

5. Is conjunction a binary operation on the set of all true statements? That is, is the conjunction of any two true statements a true statement? Is conjunction a binary operation on the set of all false statements?

6. Repeat Exercise 5 with disjunction in place of conjunction.

7. Repeat Exercise 5 with forming of conditionals in place of conjunction.

*8. Suppose that * is a binary operation defined on a set S. If T is a collection of just some of the objects of S, then we say that T is *closed* with respect to the binary operation * if the result of operating upon any two objects in T is again in the set T. Which of the following are sets of whole numbers that are closed with respect to addition of whole numbers?

(a) The set containing the numbers 0, 1, 2, 3, 4, and 5.

(b) The set of even whole numbers.

(c) The set of odd whole numbers.

(d) The set of numbers that are multiples of 3.

(e) The set of numbers that are greater than 10.

*9. Describe a set of whole numbers that

(a) Is closed with respect to both addition and multiplication.

(b) Is closed with respect to multiplication but not with respect to addition.

(c) Is not closed with respect to either addition or multiplication.

*10. With respect to the playing card operation discussed in Example 1, if S is a set of playing cards that is closed with respect to this operation and if S contains $3\Diamond$, K♣, and $5\Diamond$, then what other cards must S contain?

2.4 Properties of a Binary Operation, I

We have made the point that addition of whole numbers can be applied to every pair of whole numbers. Of course, it makes no difference in what order the numbers to be added are given—the result of the addition operation is the same in any case. Thus, for example, when addition is applied to 4 and 6, the result is the same as when addition is applied to 6 and 4. We express this by saying that addition of whole numbers has the **commutative** property. Formally, to say that addition is commutative means that $x + y = y + x$ no matter which whole numbers x and y represent.

The operation of multiplication of whole numbers also possesses the commutative property. That is to say, no matter which whole numbers x and y represent, $x \cdot y$ and $y \cdot x$ represent the same whole number: $x \cdot y = y \cdot x$.

Some operations, like addition and multiplication, possess the commutative property, but others do not. For example, the binary operation defined on playing cards (Example 1, Section 2.3) does not possess this property. If the operation were commutative, then the $5\Diamond * 2\spadesuit$ would have to be the same card as $2\spadesuit * 5\Diamond$. But $5\Diamond * 2\spadesuit = 2\Diamond$ whereas $2\spadesuit * 5\Diamond = 5\spadesuit$.

A second property of a binary operation is called the **associative** property. To introduce this property, let us consider the following four symbols:

$$8 - 4 - 2$$
$$8 + 4 + 2$$
$$8 \cdot 4 \cdot 2$$
$$8 \div 4 \div 2.$$

Now, unless we use some special rules to indicate the order in which the operations are to be performed, two of these symbols are ambiguous in the sense that the numbers those symbols represent will depend upon the order in which the two operations are performed. Try to decide which two of these symbols are ambiguous and which two are unambiguous.

The middle two symbols are meaningful, and the first and last are ambiguous. The first symbol is ambiguous because the meaning of this symbol depends upon which of the two subtractions is performed first. If the leftmost subtraction is performed first, then we get

$$8 - 4 - 2 = (8 - 4) - 2 = 4 - 2 = 2,$$

whereas if the rightmost subtraction is performed first, we get

$$8 - 4 - 2 = 8 - (4 - 2) = 8 - 2 = 6.$$

Hence the first symbol represents either 2 or 6, depending upon which of the two subtractions is performed first. Similarly, the last symbol means either

$$(8 \div 4) \div 2 = 2 \div 2 = 1$$

or

$$8 \div (4 \div 2) = 8 \div 2 = 4,$$

depending upon which division is performed first. Because of these ambiguities we do not use such symbols without parentheses. But the middle two symbols are not ambiguous; they each present a single whole number and so no parentheses are needed. For example, consider the symbol $8 + 4 + 2$. If we first perform the left most addition, we get

$$(8 + 4) + 2 = 12 + 2 = 14,$$

and we get the same result if we first perform the rightmost addition,

$$8 + (4 + 2) = 8 + 6 = 14.$$

In a similar way you can show that the symbol $8 \cdot 4 \cdot 2$ represents the single whole number 64 regardless of which of the two multiplications is performed first.

We express the fact that symbols of the form $x + y + z$ (where x, y, and z are whole numbers) are unambiguous by saying that addition of whole numbers is associative. The associative property of addition of whole numbers is usually stated in the following way: To say that addition of whole numbers is associative means that $(x + y) + z = x + (y + z)$ no matter which whole numbers x, y, and z represent. Multiplication of whole numbers is also associative. That is, $(x \cdot y) \cdot z = x \cdot (y \cdot z)$ no matter which whole numbers x, y, and z represent.

The fact that the first and last of the four symbols were ambiguous is expressed by saying that subtraction and division are not associative.

Exercises 2.4

1. Each of the following equations is justified by using either the commutative or associative property of addition. Which is used?
   (a) $6 + (3 + 2) = 6 + (2 + 3)$.   (b) $6 + (3 + 2) = (3 + 2) + 6$.
   (c) $6 + (3 + 2) = (6 + 3) + 2$.
2. Justify each equation using either the associative or the commutative property of multiplication.
   (a) $(6 \cdot 3) \cdot 2 = 6 \cdot (3 \cdot 2)$.   (b) $6 \cdot (3 \cdot 2) = (3 \cdot 2) \cdot 6$.
   (c) $6 \cdot (3 \cdot 2) = 6 \cdot (2 \cdot 3)$.

3. Justify each of the following equations by using the commutative and/or associative properties of addition and/or multiplication.
   (a) $6 \cdot (3 \cdot 2) = (3 \cdot 6) \cdot 2$.   (b) $6 + (2 + 3) = (2 + 6) + 3$.
   (c) $6 + (3 + 2) = (6 + 2) + 3$.   (d) $6 + (3 + 2) = (2 + 6) + 3$.
4. Each of these symbols is ambiguous. What are the different possible meanings of each symbol?
   (a) $2 + 3 \cdot 4$.   (b) $6 - 2 + 2$.
   (c) $6 \cdot 3 - 2$.
   (d) $20 - 8 - 6 - 2$. (*Hint:* There are five different ways to interpret this symbol.)
5. Is the conjunction operation commutative? That is, if p and q are statements, are the statements $p \wedge q$ and $q \wedge p$ logically equivalent? What about the disjunction operation?
6. Is the conjunction operation associative? That is, if p, q, and r are statements, are the compound statements $(p \wedge q) \wedge r$ and $p \wedge (q \wedge r)$ logically equivalent? What about the disjunction operation?
7. Give counterexamples to the commutativity and associativity of the conditional operation. That is, find statements p and q such that $p \rightarrow q$ and $q \rightarrow p$ are not equivalent and such that $(p \rightarrow q) \rightarrow r$ and $p \rightarrow (q \rightarrow r)$ are not equivalent.
8. Look up the words *commutative* and *associative* in the dictionary and see if you can understand why these properties were named as they were.
9. Use the cards 3♡, 6◇, and J♠ to demonstrate that the operation on playing cards (Example 1, Section 2.3) is associative. Then repeat the demonstration using three other cards of your own selection.
*10. (Before you begin this problem, recall how an addition table can be used to find the sum of one-digit numbers.) It is possible to define binary operations by means of tables. In Table 2.1 we define a binary operation (which we shall denote by *) on the set of letters a, b, c, and d. To find the result of applying this operation to a pair of letters x and y (that is, to find the object $x * y$), find the letter in *row x* and in *column y*. For example, $c * b = d$ and $a * d = d$. Is the operation defined by this table commutative? How can you tell that the operation is commutative simply by observing the geometric properties of the four-by-four array of letters in Table 2.1?

Table 2.1

	Columns			
*	a	b	c	d
a	a	b	c	d
b	b	c	d	a
c	c	d	a	b
d	d	a	b	c

Rows

2.5 Properties of a Binary Operation, II

In this section we shall continue the discussion of properties of a binary operation that we began in the last section.

The whole number 0 plays a distinguished role relative to addition of whole numbers. Zero is the only number that, when added to a number, gives that number as the sum. That is, no matter which whole number x represents, $0 + x = x$ and $x + 0 = x$. Is there a number that acts toward multiplication in the same way that 0 acts toward addition? Surely, because $1 \cdot x = x$ and $x \cdot 1 = x$, 1 is the number we want. Because of these special properties of 0 and 1, we call the whole number 0 the **additive identity** and we call 1 the **multiplicative identity.** See Figure 2.5.

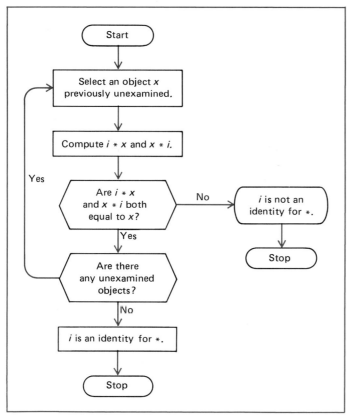

Figure 2.5 *Flow chart for determining whether an object i is an identity for a binary operation * defined on a finite set.*

The number 0 acts very differently toward multiplication than it does toward addition. Indeed, the product of 0 with any whole number at all is 0: $0 \cdot x = 0$ and $x \cdot 0 = 0$ for all whole numbers x. Because of the way 0 behaves with respect to multiplication, we call 0 the **annihilator** for multiplication.

Example 1.   The binary operation of addition defined on the set of even whole numbers has 0 as an identity. But multiplication defined on the set of even whole numbers has no identity because the only possible candidate for such an identity element, 1, is not an even whole number.

Example 2.   With reference to the binary operation on playing cards (Example 1, Section 2.3), the card 2◇ is not an identity for this operation. To prove this, we need only demonstrate by means of an example that there is a card x such that $x * 2◇ \neq x$. For instance, 3♣ * 2◇ ≠ 3♣. Is there any card that is an identity for this operation?

Suppose we know that

$$x + 4 = y + 4,$$

where x and y are whole numbers and $+$ denotes addition of whole numbers. What can you say about x and y? Surely, x and y are equal: $x = y$. We have been able to "erase" the 4 from both sides of the first equation to obtain the second. Because each of the equations $x + z = y + z$ and $z + x = z + y$ implies that $x = y$, we say that addition of whole numbers possesses the **cancellation** property.

Multiplication of whole numbers also possesses a cancellation property. If x, y, and z are whole numbers such that either $xz = yz$ or $zx = zy$ and if z *is not equal to* 0, then we may conclude that $x = y$. This cancellation property says that we can cancel any whole number except 0. Why can we not cancel 0? Consider the equation $3 \cdot 0 = 7 \cdot 0$. This equation is a true statement because any whole number multiplied by 0 is 0. But we cannot cancel the 0 from both sides of this equation because it is obviously untrue that $3 = 7$. Hence we cannot expect to be able to cancel 0 with respect to multiplication.

Here is an example that shows how these cancellation properties may be used to help solve equations.

Example 3.   Solve the equation $5x = 2x + 6$.

Solution: Since $5x = 2x + 3x$, we can rewrite this equation in the form

$$2x + 3x = 2x + 6.$$

Applying the cancellation property of addition, we get the equation

$$3x = 6.$$

On rewriting 6 as $3 \cdot 2$, we get the equation

$$3x = 3 \cdot 2.$$

Now, applying the cancellation property of multiplication (since 3 is not zero), we finally obtain

$$x = 2.$$

Hence the solution of the given equation is 2.

The last property we shall discuss at this time is called the **distributive** property. This is not a property that is possessed by one operation alone; it is a property that is shared by two operations. For example, consider the operations of addition and multiplication of whole numbers. The symbol $3 \cdot (4 + 5)$ involves both of these operations and may be simplified in two different ways. One way is to perform the addition first and write

$$3 \cdot (4 + 5) = 3 \cdot 9 = 27.$$

The other way is to multiply 3 and 4, multiply 3 and 5, and then add these products:

$$3 \cdot (4 + 5) = (3 \cdot 4) + (3 \cdot 5) = 12 + 15 = 27.$$

This second computation involves the use of the distributive property of multiplication and addition. Formally, this property is stated as follows: To say that addition and multiplication of whole numbers share the distributive property means that $x \cdot (y + z) = (x \cdot y) + (x \cdot z)$ no matter which whole numbers x, y, and z represent. Here are some examples of how this property is used.

Example 4. The product of 34 and 57 is generally computed as follows:

$$\begin{array}{r} 57 \\ 34 \\ \hline 228 \\ 171 \\ \hline 1938 \end{array}$$

Now examine the following equation, which utilizes the distributive property:

$$57 \cdot 34 = 57 \cdot (30 + 4)$$

$$= (57 \cdot 30) + (57 \cdot 4)$$

$$= 1710 + 228$$

$$= 1938$$

Comparing the "vertical" method for finding this product (this method is called the *ordinary multiplication algorithm*[1]) with the "horizontal" method, we see that the ordinary "vertical" algorithm uses the distributive property and uses this property in a very important way. Indeed, the distributive property can perhaps be regarded as the most important aspect of this "vertical" sort of algorithm.

The statement

$$a(b + c) = ab + ac,$$

which we have called the distributive property for addition and multiplication, is more properly called the **left distributive** property for addition and multiplication. The **right distributive** property for these two operations is

$$(b + c)a = ba + ca.$$

This right distributive property can be obtained from the left distributive property by using the commutative property of multiplication.

Addition and multiplication are not the only operations that share distributive properties of one kind or another. We shall see examples of other distributive properties in our later work. See also Exercise 12.

Exercises 2.5 ——————————————————

1. Explain why the card $5\diamondsuit$ is not an identity for the playing card operation defined in Example 1, Section 2.3.
2. Does the operation of conjunction have an identity? That is, is there a statement i such that $p \wedge i \longleftrightarrow i \wedge p \longleftrightarrow p$ no matter which statement p represents?
3. Does the operation of disjunction have an identity? That is, is there a statement i such that $p \vee i \longleftrightarrow i \vee p \longleftrightarrow p$ no matter which statement p represents?
4. Which of the following cancellations are correct and which are the result of

——————————————————

[1] By an algorithm we mean a computational method by means of which we determine sums, products, square roots, and so forth.

——————————————————

a misuse of the real cancellation properties? (All letters represent whole numbers.)

(a) $3x + 7 = 3x + 9$ implies $x + 7 = x + 9$, since we can cancel the 3's from both sides of an equation.

(b) $7 + x = 3(a + x)$ implies that $7 = 3a$, since we can cancel the x from both sides of the equation.

(c) $6a + b = 6x + c$ implies that $a + b = x + c$, since we can cancel the 6 from both sides of the equation.

(d) $5x + 9 = 5y + 9$ implies $x = y$, since we can cancel first 9 then 5 from both sides of the equation.

(e) $(a - b)(c + b) = (a - b)(2c)$ implies $c + b = 2c$.

(f) $(a + b)(c + b) = (a + b)(2c)$ implies $c + b = 2c$.

(g) $(a + b + 1)(c + b) = (a + b + 1)(2c)$ implies $c + b = 2c$.

5. Use the cancellation properties of addition and/or multiplication to solve the following equations.

(a) $x + 2 = 5$.

(b) $4x = 32 + 2x$.

6. Explain how the distributive property is used in multiplying 34 by 18 using the ordinary (vertical) multiplication algorithm. In using this algorithm do you ever really multiply 34 by 18? What numbers do you multiply?

7. Make truth tables and verify that conjunction and disjunction share the following distributive properties:

(a) $p \land (q \lor r) \longleftrightarrow (p \land q) \lor (p \land r)$.

(b) $p \lor (q \land r) \longleftrightarrow (p \lor q) \land (p \lor r)$.

8. One of the following distributivelike properties is true and the other is false. Which is true?

(a) $p \land (q \rightarrow r) \longleftrightarrow (p \land q) \rightarrow (p \land r)$.

(b) $p \rightarrow (q \land r) \longleftrightarrow (p \rightarrow q) \land (p \rightarrow r)$.

9. Complete the justifications for each step in this argument.

Theorem.

$(a + b)(c + d) = ac + ad + bc + bd$.

Proof:

(1) $(a + b)(c + d) = a(c + d) + b(c + d)$. (1) This is an application of what distributive property for addition and multiplication?

(2) $a(c + d) + b(c + d) = [ac + ad] + b(c + d)$. (2) This is an application of what distributive property?

(3) $[ac + ad] + b(c + d) = [ac + ad] + [bc + bd]$. (3) This is an application of what distributive property?

(4) $[ac + ad] + [bc + bd] = ac + ad + bc + bd$. (4) We can omit the parentheses because of what property?

(5) $(a + b)(c + d) = ac + ad + bc + bd$. (5) This follows from equations (1) through (4) by repeated use of what property of equality?

10. Use Exercise 9 to complete these equations.
    (a) $(n + m)(n + m) =$ (b) $(x + y)(2 + ab) =$
11. We can use Exercise 9 to show that $(x + 2)^2 = x^2 + 4x + 4$. The Greeks proved this geometrically by using the accompanying figure. Explain how they did this.

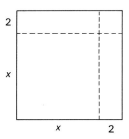

12. Each of the following is a distributivelike statement involving two arithmetic operations, but not all of them are true. Which are false?
    (a) $a \cdot (b - c) = (a \cdot b) - (a \cdot c)$.
    (b) $(a - b) \cdot c = (a \cdot c) - (b \cdot c)$.
    (c) $(a \div b) + c = (a \div c) + (b \div c)$.
    (d) $a + (b \div c) = (a + b) \div (a + c)$.
    (e) $(a + b) \div c = (a \div c) + (b \div c)$.
    (f) $(a - b) \div c = (a \div c) - (b \div c)$.
    (g) $(a + b) - c = (a - c) + (b - c)$.
    (h) $(a \cdot b) \div c = (a \div c) \cdot (b \div c)$.
13. The flow chart in Figure 2.5 works only provided the set upon which the binary operation * is defined is a finite set. Explain.
14. Make a flow chart that will show how to determine whether an object $\mathcal{O}$ is an annihilator for a binary operation * defined on a finite set of objects.

2.6 Arithmetic Modulo Six, I

We have been discussing the idea of a mathematical system largely with reference to the mathematical system of whole numbers. To help you to see the essential aspects of a mathematical system more clearly, we shall now introduce a new and only partially familiar mathematical system called *arithmetic modulo six*.

The mathematical system called arithmetic modulo six involves only six objects, the whole numbers 0 through 5. Let us arrange these six numbers around the circumference of a circle, as shown in Figure 2.6. The only relation that we shall need is equality of objects. When we write $n = m$ (where n and m represent objects of this new system), we shall mean simply that the symbols n and m represent

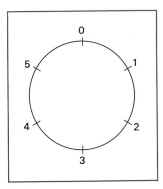

Figure 2.6

the same object of the system. So let us go on to the binary operations of this system. First, let us define an operation that derives from addition of whole numbers. For convenience, we shall call this operation 6-addition to remind us that it resembles addition of whole numbers but is defined only on the six objects of this new system. Here is how 6-addition is performed. Given a pair of objects, x and y, start at the number x on the circle (Figure 2.6) and move y distances in the clockwise direction. The number at which you stop is the 6-sum of x and y. We shall denote 6-addition by the symbol $\oplus$. For example (and you should verify each of these statements),

$$5 \oplus 3 = 2, \qquad 3 \oplus 3 = 0, \qquad 4 \oplus 4 = 2, \qquad \text{and} \qquad 5 \oplus 2 = 1.$$

Table 2.2 is a 6-addition table for this operation. The entry in row n and column m represents the 6-sum of the objects n and m (in that order). Hence the entry in row n and column m is the object $n \oplus m$.

Example 1. The equation $x \oplus 4 = 2$ is solved by finding those objects that give 2 when 4 is 6-added to them. From Table 2.2 we see that $4 \oplus 4$ is equal to 2

Table 2.2 6-Addition Table for Arithmetic Modulo Six

		Columns					
	$\oplus$	**0**	**1**	**2**	**3**	**4**	**5**
	0	0	1	2	3	4	5
	1	1	2	3	4	5	0
Rows	**2**	2	3	4	5	0	1
	3	3	4	5	0	1	2
	4	4	5	0	1	2	3
	5	5	0	1	2	3	4

and that no other object plus 4 is equal to 2. Hence the only solution of this equation is 4.

Example 2. Solve the equation $x \oplus x = 2$.

Solution: Examining the 6-addition table (Table 2.2), we see that $1 \oplus 1 = 2$ and $4 \oplus 4 = 2$. Hence 1 and 4 are solutions. There are no other solutions.

Example 3. Is there an identity for 6-addition?

Solution: Yes, 0 is the identity. From Table 2.2 we see that $x \oplus 0 = x$ and $0 \oplus x = x$ for all $x = 0, 1, 2, 3, 4,$ and 5.

If by moving around the circle in a clockwise direction we perform 6-addition, then moving around the circle in a counterclockwise direction ought to define 6-subtraction. Let us define 6-subtraction as follows: Given two objects x and y, to 6-subtract y from x start at the point x on the circle and move y distances in the counterclockwise direction. The landing point is the result of having 6-subtracted y from x. Let us denote 6-subtraction by $\ominus$.

Example 4. $3 \ominus 5 = ?$

Solution: Starting at 3 and moving 5 distances in the counterclockwise direction lands us on 4. Hence $3 \ominus 5 = 4$.

Example 5. Find the solution of the equation $4 \ominus x = 5$.

Solution: If we start at 4, how many distances must we move in the counterclockwise direction in order to land on 5? The answer is 5 distances, and therefore the solution of $4 \ominus x = 5$ is 5.

Example 6. $(x \oplus y) \ominus y = ?$ $(x \ominus y) \oplus y = ?$

Solution: $(x \oplus y) \ominus y = x$, since starting at x, moving y distances clockwise, and then moving y distances counterclockwise lands us back at x. Similarly, $(x \ominus y) \oplus y$ is the result of starting at x, moving y distances in the counterclockwise direction, and then moving y distances in the clockwise direction. Thus $(x \ominus y) \oplus y = x$.

Example 6 demonstrates an important connection between 6-addition and 6-subtraction. We see that 6-subtracting a number "undoes" the effect of having 6-added that number and that 6-adding a number "undoes" the effect of having 6-subtracted that same number. Because of this connection between 6-addition and 6-subtraction, we say that these two operations are **inverse operations.** More generally, when two binary operations have the effect of "undoing" each other in this way, we say that each operation is the inverse of the other.

By this time you should be wondering what is so special about the number 6. Actually, there is nothing special at all about 6. There is a mathematical system similar to arithmetic modulo six built upon each of the whole numbers greater than one. For example, **arithmetic modulo four** is the mathematical system built upon the four objects 0, 1, 2, and 3. An operation called 4-addition can be defined upon these objects using a "number circle" in the same general way that 6-addition was defined here. We shall consider this mathematical system in the exercises.

In the next section we shall continue our discussion of arithmetic modulo six by introducing an operation called 6-multiplication.

Exercises 2.6

1. Find the 6-sum of the objects 3, 5, 4, 2, 1, 3, and 5.
2. Use the objects 3, 4, and 5 to demonstrate that 6-addition is associative. Is 6-addition commutative?
3. Solve the following equations in arithmetic modulo six. Remember that some of them may have more than one solution.
   (a) $x \ominus 2 = 4$.   (b) $3 \ominus (x \oplus x) = 1$.   (c) $5 \oplus (x \ominus 2) = 0$.
4. Is 6-subtraction a binary operation? That is, is it true that no matter which objects x and y represent, $x \ominus y$ always represents a unique object? Construct a table for 6-subtraction similar to the 6-addition table shown in Table 2.2. (The object entered in row x and column y will be $x \ominus y$.)
5. Verify by means of examples that 6-subtraction is neither commutative nor associative.
6. If x is an object in arithmetic modulo six, then by the **additive inverse** of x we mean that object y such that $x \oplus y = 0$. Find the additive inverse of each object in arithmetic modulo six. (For example, 4 is the additive inverse of 2 since $2 \oplus 4 = 0$.
7. Does 6-addition have the cancellation property? That is, if $a \oplus b = c \oplus b$ can you always conclude that $a = c$? (You can answer this question in terms of the number circle.)
8. It would appear reasonable (on the basis of what we learned about the relation *is less than* in the system of whole numbers) to say that in arithmetic modulo six an object x is less than an object y if there is a nonzero object d such that $x \oplus d = y$. Show that if we tried to use such a definition, we would be able to conclude both that $2 < 3$ and that $3 < 2$. We conclude, therefore, that it is impossible to define a *less than* relation in this number system.
9. Using the numbers 0, 1, 2, 3, and 4 and using a number circle (similar to the one in Figure 2.6), we can build a mathematical system called **arithmetic modulo five.** In this system, $2 \oplus 4 = 1$, since starting at 2 on the number circle for this system and moving four distances around the circle lands us on 1. Construct a table for 5-addition in arithmetic modulo five.

10. In arithmetic modulo five,
    (a) Is 0 an identity for 5-addition?
    (b) Is 5-addition commutative and associative?
    (c) What is the 5-sum of 4, 3, 2, 4, and 1?
    (d) What is the solution of the equation $x \ominus 2 = 3$?
11. Using the numbers 0, 1, 2, and 3 placed around the circumference of a circle, we can define an operation called 4-addition and a mathematical system called **arithmetic modulo four.** Construct a 4-addition table for this system. Then use this table to solve the following equations.
    (a) $x \oplus x = 3.$ (b) $x \oplus x = 2.$
    (c) $x \oplus x = 0.$ (d) $x \oplus 2 = 1.$
    (e) $x \ominus 3 = 2.$ (f) $3 \ominus x = 1.$

2.7 Arithmetic Modulo Six, II

Let x and y represent two objects in arithmetic modulo six. We have seen how to define additionlike and subtractionlike operations. How can we define a multiplicationlike operation? One way to define such an operation (denoted by $\otimes$) is to regard 6-multiplication as repeated 6-addition. Thus, for example, we would find $5 \otimes 3$ as follows:

$$5 \otimes 3 = 5 \oplus 5 \oplus 5$$
$$= 4 \oplus 5$$
$$= 3.$$

Hence $5 \otimes 3 = 3$ in arithmetic modulo six.

After having performed a great many 6-multiplications, we might discover an easier way to 6-multiply. Consider the problem $5 \otimes 3 = ?$ again. If we multiply these two numbers in the mathematical system of whole numbers (using the operation of ordinary multiplication of whole numbers), we get 15 as the product. If we subtract (or "cast out") as many 6's as possible from 15, we are left with 3: $15 - 6 = 9, 9 - 6 = 3$. Thus we see that the 6-product of 5 and 3 is obtained by casting out as many 6's as possible from the product 15. This is a general method for computing 6-products. See Figure 2.7. Thus to find the 6-product of 5 and 5, multiply 5 and 5 using multiplication of whole numbers and then cast out as many 6's as possible: $25 - 6 = 19, 19 - 6 = 13, 13 - 6 = 7, 7 - 6 = 1$. Hence $5 \otimes 5 = 1$. Using repeating 6-addition, we can compute this 6-product as follows:

$$5 \otimes 5 = 5 \oplus 5 \oplus 5 \oplus 5 \oplus 5$$
$$= 4 \oplus 5 \oplus 5 \oplus 5$$
$$= 3 \oplus 5 \oplus 5$$
$$= 2 \oplus 5$$
$$= 1.$$

The same kind of casting-out procedure can be used for 6-addition.

Example 1.   Find the 6-sum and 6-product of 3 and 4 by casting out 6's.

Solution: $3 + 4 = 7$ in the system of whole numbers. Casting out a 6 from 7, we get 1. Hence $3 \oplus 4 = 1$. The product of 3 and 4 in the system of whole numbers is 12. We can cast out two 6's from 12, leaving 0. Hence $3 \otimes 4 = 0$ in arithmetic modulo six.

Table 2.3 is a partially filled-in 6-multiplication table. Before going further, you should complete this table.

Before we leave arithmetic modulo six, let us comment upon some of the essential differences between this arithmetic and the usual arithmetic of whole numbers. In the system of whole numbers, it is not possible to find two *nonzero* objects whose product is zero. But in arithmetic modulo six there do exist nonzero objects whose

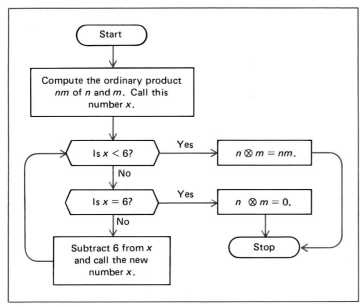

Figure 2.7 *Flow chart for computing $n \otimes m$.*

Table 2.3 6-Multiplication
Table for Arithmetic Modulo Six

				Columns		
$\otimes$	**0**	**1**	**2**	**3**	**4**	**5**
0	0	0	0	0		
1	0	1	2	3		
2	0	2	4	0	2	4
3						
4				0	4	2
5			4		2	

Rows

6-product is zero. For example, $3 \otimes 4 = 0$. In the system of whole numbers, subtraction is not a binary operation (for example, $5 - 7$ has no meaning in that system). In arithmetic modulo six, 6-subtraction is a binary operation because $x \ominus y$ has a unique meaning no matter which of the six objects x and y represent. (To see why this is true, think about how subtraction is defined in terms of moving around the circle.) In the system of whole numbers, multiplication has the cancellation property (for nonzero numbers). In arithmetic modulo six, 6-multiplication does not have this property. For example, from the equation $4 \otimes 3 = 2 \otimes 3$, we cannot cancel the 3 and conclude that $4 = 2$.

Finally, let us comment on why we have not tried to define a divisionlike operation. Suppose we attempted to define a process called 6-division. Since (from the 6-multiplication table)

$$3 = 3 \otimes 1 \qquad \text{and} \qquad 3 = 3 \otimes 3,$$

we would have to conclude (if we tried to define 6-division) that

$$3 \oplus 3 = 1 \qquad \text{and} \qquad 3 \oplus 3 = 3.$$

That is, the result of 6-dividing 3 by 3 would be equal to two different objects—this "quotient" would not be a single number. This is sufficiently unpleasant so that we do not define 6-division at all. Generally speaking, mathematics does not deal with processes that do not yield unique results. There are exceptions to this, but they are isolated and are of no interest to us.

Exercises 2.7 _____

1. Find the 6-product of the objects 4, 4, 5, 2, and 5.
2. Solve the following equations in arithmetic modulo six.
   (a) $x \otimes x = 1$.
   (b) $x \otimes x = 4$.
   (c) $(2 \otimes x) \ominus 3 = 0$.
   (d) $(2 \otimes x) \ominus 2 = 0$.
   (e) $x \otimes (3 \otimes 5) = 1$.
   (f) $(x \ominus 1) \otimes (x \ominus 2) = 0$.

3. Show that 6-multiplication does not possess the cancellation property by finding objects a, b, and c such that $a \otimes c = b \otimes c$ but $a \neq b$.

4. Use the objects 2, 3, and 5 to demonstrate that 6-addition and 6-multiplication share the right and left distributive properties.

5. Use the objects 3, 4, and 5 to demonstrate that 6-multiplication is associative. Is the operation commutative? Is there an identity for 6-multiplication?

*6. If x is an object in arithmetic modulo six, then we call an object y the **multiplicative inverse** of x if $x \otimes y = 1$. Does every object in arithmetic modulo six possess a multiplicative inverse?

7. Write out the table for 5-multiplication in arithmetic modulo five. (You can find the entries for this table by first multiplying the objects as if they were whole numbers and by then casting out as many fives as possible from this product.)

8. We saw that in arithmetic modulo six it was possible to find nonzero objects whose 6-product is zero. Can nonzero objects be found in arithmetic modulo five whose 5-product is zero? (Refer to the table you constructed in Exercise 7.)

9. Does every nonzero object in arithmetic modulo five have a multiplicative inverse? (Refer to Exercise 6.)

10. If we try to define 5-division in arithmetic modulo five, we have better luck than we did when we tried to define 6-division in arithmetic modulo six. Show that every object can be 5-divided by every nonzero object in arithmetic modulo five. Make a 5-division table (which, of course, will not have any column corresponding to 0, since we cannot divide by zero).

11. A factory is working 24 hours a day to produce a certain piece of machinery. It takes 3 hours to produce one of these products; starting at 9 A.M. on a certain day, they begin to fill an order for 34 of these pieces of machinery. At what hour of the day will they finish the order? What modular arithmetic system is involved in this problem?

12. Make a flow chart for 6-addition similar to the flow chart for 6-multiplication in Figure 2.7.

13. Make a flow chart for performing 6-subtraction by the method analogous to that shown for 6-multiplication in Figure 2.7.

3

The System of Whole Numbers

We have already discussed many of the most fundamental properties of the mathematical system of whole numbers. In this chapter we are going to pursue some of the consequences of these fundamental properties. The first half of the chapter consists of a formal study of some of these consequences, including definitions of subtraction and division. The second half of the chapter consists of a less formal introduction to some of the ideas of elementary number theory.

3.1 Review of Basic Properties

We have already discussed most of the basic properties of the system of whole numbers that are used in elementary arithmetic involving whole numbers. So let us begin by listing these properties.

1. The binary operation of addition of whole numbers has the following properties:
   (A) Addition is commutative: $a + b = b + a$ for all whole numbers a and b.

(B) Addition is associative: $a + (b + c) = (a + b) + c$ for all whole numbers a, b, and c.

(C) Addition has an additive identity (0): $a + 0 = a$ and $0 + a = a$ for all whole numbers a.

(D) Addition has the cancellation property: $a + b = c + b$ implies that $a = c$.

2. The binary operation of multiplication of whole numbers has the following properties:

(A) Multiplication is commutative: $a \cdot b = b \cdot a$, for all whole numbers a and b.

(B) Multiplication is associative: $a \cdot (b \cdot c) = (a \cdot b) \cdot c$, for all whole numbers a, b, and c.

(C) Multiplication has an identity (1): $x \cdot 1 = x$ and $1 \cdot x = x$ for all whole numbers x.

(D) Multiplication has the cancellation property for nonzero whole numbers: $a \cdot b = c \cdot b$ implies $a = c$ if b is not zero.

3. Multiplication and addition share both right and left distributive properties: $a \cdot (b + c) = (a \cdot b) + (a \cdot c)$ and $(b + c) \cdot a = (b \cdot a) + (c \cdot a)$, for all whole numbers a, b, and c.

In the next three sections we are going to undertake a program of development of the system of whole numbers based upon these fundamental properties.

3.2 Selected Consequences, I

Let us begin our development of the system of whole numbers from the basic properties of the last section by defining the inequality relations and by stating a few theorems concerning these relations. We shall begin with the relation **is less than.**

Definition of *Is Less Than*.
If x and y are whole numbers, then to say that x is less than y, in symbols $x < y$, means that there is some nonzero whole number d such that $x + d = y$.

The definition states that x is less than y if, when x is increased by some nonzero amount d, the sum is y. For example, 5 is less than 8, $5 < 8$, because when 5 is increased by 3 the resulting sum is 8. On the other hand, $6 \nless 5$ because there is no nonzero whole number d such that $6 + d = 5$.

The most important theorems about the *is less than* relation are the following. These theorems relate this relation to the operations of addition and multiplication.

Theorem A.

If x, y, and z are whole numbers and if $x < y$, then $x + z < y + z$.

Theorem B.

If x, y, and z are whole numbers, if $x < y$, and if $z \neq 0$, then $xz < yz$.

Theorem A states that the same whole number can be added to both sides of an inequality involving the *is less than* relation, and Theorem B states that both sides of such an inequality can be multiplied by the same nonzero whole number. Let us prove Theorem A. This proof is typical of a *direct proof* in that the statements that comprise the argument lead directly from the hypothesis of the theorem to its conclusion. In order to make these statements and the logical connections between them abundantly obvious, we have arranged them in vertical form. Here is the proof of Theorem A:

Hypothesis: $x < y$.
Conclusion: $x + z < y + z$.

Proof:

Statements	Justifications
1. $x < y$.	1. Hypothesis.
2. There is a nonzero whole number d such that $x + d = y$.	2. From (1) by using the definition of *is less than*.
3. $(x + d) + z = y + z$.	3. By adding z to both sides of the equation in (2).
4. $x + (d + z) = y + z$.	4. From (3) by using the associativity of addition.
5. $x + (z + d) = y + z$.	5. From (4) by using the commutativity of addition.
6. $(x + z) + d = y + z$.	6. From (5) by using the associativity of addition.
7. $x + z < y + z$.	7. This follows from (6) by using the fact that d is nonzero and the definition of *is less than*.

This completes the proof.

We shall supply statements leading to a proof of Theorem B in the exercises. The second important relation is the relation **is greater than.**

Definition of *Is Greater Than.*

When we say that x is greater than y, $x > y$, we mean that y is less than x.

According to this definition, then, $x > y$ if and only if $y < x$. There are two other inequality relations of interest, each of which is constructed using one of these two inequality relations together with the basic concept of equality. We say that x *is less than or equal to* y, $x \leq y$, if either $x < y$ or $x = y$. Hence, for example, $2 \nleq 1$, $2 \leq 2$, and $2 \leq 3$. The relation *is greater than or equal to* is defined similarly.

Theorems A and B have analogs for these other three inequality relations. You should write out a few of these analogs.

Exercises 3.2

1. Use the definitions of *is less than* and *is greater than* to justify each of the following statements.
   (a) $3 < 5$.   (b) $5 > 3$.   (c) $0 < 5$.   (d) $5 > 0$.
2. State a definition for the relation *is greater than or equal to.*
3. Indicate the duplications in the list: $<$, $\leq$, $>$, $\geq$, $\nless$, $\nleq$, $\ngtr$, and $\ngeq$.
4. Which of the three properties of a relation are possessed by each of the four inequality relations $<$, $>$, $\leq$, and $\geq$?
5. State the analog of Theorem A for the relation $>$. State the analog of Theorem B for the relation $\geq$.
6. Supply justifications for the statements of this proof of Theorem B.

Hypothesis: $x < y$ *and* $z \neq 0$.
Conclusion: $xz < yz$.

Proof:
1. $x < y$.
2. $x + d = y$ for some nonzero whole number d.
3. $(x + d)z = yz$.
4. $xz + dz = yz$.
5. $dz > 0$.
6. $xz < yz$.

7. Why must we restrict z from being equal to 0 in Theorem B?
8. If you took a friend to dinner in an unfamiliar restaurant and if the bill were larger than you had anticipated and if you were not sure how much money you actually had with you, which of the two relations $<$ and $\leq$ would you be most immediately interested in?
9. Here is a proof that the relation *is less than* has the transitive property. Supply the justifications.

Theorem.
If $x < y$ *and* $y < z$, *then* $x < z$.
Hypotheses: $x < y$ *and* $y < z$.
Conclusion: $x < z$.

Proof:
1. $x < y$ and $y < z$.
2. $x + d = y$ and $y + e = z$, for some nonzero whole numbers d and e.

3. $(x + d) + e = z.$
4. $x + (d + e) = z.$
5. $d + e$ is nonzero.
6. $x < z.$

3.3 Selected Consequences, II

As the second step in our development of the system of whole numbers, we shall define subtraction and division of whole numbers and state a few theorems relating to these processes.

First, let us define subtraction. Observe how addition is used in this definition.

Definition of Subtraction.
If x and y are whole numbers such that $x \geq y$, then $x - y$ represents that whole number d such that $x = y + d$. If $x < y$, then the symbol $x - y$ is meaningless in the system of whole numbers.

This definition tells us that the equations

$$x - y = d \qquad \text{and} \qquad x = y + d$$

express exactly the same relationship between the numbers x, y, and d. For example, because $6 = 4 + 2$ we know, according to this definition, that $6 - 4 = 2$. The difference $8 - 5$ is equal to 3 because $8 = 5 + 3$. But the symbol $7 - 9$ is meaningless (in this mathematical system) because there is no whole number d such that $7 = 9 + d$.

Subtraction is important in the system of whole numbers because addition and subtraction are inverse operations. By this we mean that each of these processes "undoes" the result of having performed the other. More precisely, we have the following two theorems:

Theorem C.
If x and y are whole numbers, then $(x + y) - y = x$.

Theorem D.
If x and y are whole numbers and if $x \geq y$, then $(x - y) + y = x$.

Here are the proofs of these two theorems.

Proof of Theorem C:

Let x and y be whole numbers. Then, because addition is commutative, we know that

$$x + y = y + x.$$

Now regard the left-hand member of this equation as a single entity and think of the right side as the sum of two numbers. Then, according to the definition of subtraction, this equation expresses exactly the same idea as does the equation

$$(x + y) - y = x.$$

This is just what we wanted to prove.

Proof of Theorem D:

Let x and y be whole numbers such that $x \geq y$. Since $x \geq y$, the symbol $x - y$ is meaningful in this system and so we can write

$$x - y = x - y.$$

Regard the right-hand side of this equation as a single number and the left-hand side as being the difference of two numbers. Then the definition of subtraction tells us that

$$x = y + (x - y).$$

Using commutativity of addition, we can rewrite this equation in the form

$$x = (x - y) + y,$$

and this can be rewritten as

$$(x - y) + y = x,$$

which is exactly what we wanted to prove.

There are many theorems having to do with subtraction that we could prove now, but let us go on and define division of whole numbers.

Definition of Division.

If x and y are whole numbers and if y is not equal to zero, then $x \div y$ (the result of dividing x by y) is that number q such that $x = y \cdot q$, if there is such a number. If no such number q exists, then the symbol $x \div y$ is meaningless and x cannot be divided by y.

The definition of division tells us that the equations

$$x \div y = q \qquad \text{and} \qquad x = y \cdot q$$

express precisely the same relationship between the numbers x, y, and q. Observe how the definition of division in terms of multiplication resembles the definition of subtraction in terms of addition.

Example 1. The symbol $8 \div 4$ represents the whole number 2 because it is true that $8 = 4 \cdot 2$. Since $15 = 5 \cdot 3$, we may conclude that $15 \div 5 = 3$. The symbol $15 \div 7$ has no meaning in the system of whole numbers because there is no whole number q such that $15 = 7 \cdot q$.

In this definition of division of whole numbers, we specifically restricted y from being 0. That is, according to our definition, it is not permissible to divide by 0. Why did we say this? There are two cases to consider: first, why we cannot divide a nonzero whole number by 0, and second, why we cannot divide 0 by 0. The reasons are different in each case.

To see why we cannot divide a nonzero whole number by 0, let us consider the illustration of 5 divided by 0. If it were possible to divide 5 by 0 (that is, if the symbol $5 \div 0$ had any meaning), then we could write $5 \div 0 = q$, where q is some definite whole number. Then, according to our definition, because $5 \div 0 = q$, it would be true that $5 = 0 \cdot q$. But $0 \cdot q = 0$, and so we would have $5 = 0$, which is absurd. Hence $5 \div 0$ cannot have any meaning at all.

To explain why $0 \div 0$ has no meaning requires different reasoning. First, we note that each of the following equations is true: $0 = 0 \cdot 0$, $0 = 0 \cdot 1$, $0 = 0 \cdot 2$, and so on. According to our definition, because $0 = 0 \cdot 0$, it would have to be true that $0 \div 0 = 0$. Also, since $0 = 0 \cdot 1$, it would have to be true that $0 \div 0 = 1$. Since $0 = 0 \cdot 2$, it would have to be true that $0 \div 0 = 2$, and so on. Thus we see that $0 \div 0$ would have to be equal to 0, to 1, to 2, and to 3, and in fact to every whole number. This is entirely undesirable because we want the answer to a division problem to be unique. Hence we do not assign any meaning to $0 \div 0$.

Division and multiplication are inverses of each other in the same general way that subtraction and addition are inverses. More precisely, we have the following two theorems.

Theorem E.
If x and y are whole numbers and $y \neq 0$, then $(x \cdot y) \div y = x$.

Theorem F.
If x and y are whole numbers such that $x \div y$ is also a whole number, then $(x \div y) \cdot y = x$.

The proofs of these theorems are included in the exercises. The following examples illustrate how Theorems A through F are used in solving even the simplest equations and inequalities.

Example 2.   Solve the equation $x - 3 = 5$.

Solution: The usual way to solve this equation is to begin by adding 3 to both sides:

$$(x - 3) + 3 = 5 + 3.$$

Then, according to Theorem D, we can simplify the left-hand side of this equation and write $x = 5 + 3$ or $x = 8$.

Example 3.   Solve the equation $2x + 4 = 6$.

Solution: We would begin by subtracting 4 from both sides and obtaining the equation $(2x + 4) - 4 = 6 - 4$. Then, using Theorem C, we would rewrite this equation as $2x = 6 - 4$ so that $2x = 2$. Now $2x = x \cdot 2$, so we can write $x \cdot 2 = 2$. Dividing both sides by 2, we obtain the equation $(x \cdot 2) \div 2 = 2 \div 2$. According to Theorem E, then, we can write $x = 2 \div 2$ so that $x = 1$.

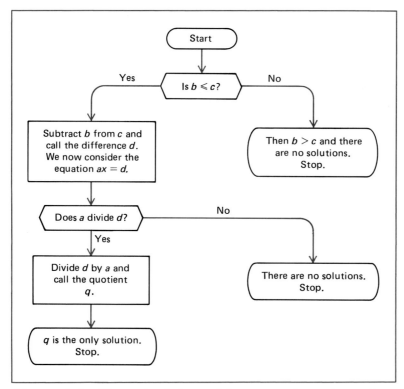

Figure 3.1 *Flow chart for solving equation of the form*
$ax + b = c$.

The flow chart in Figure 3.1 shows how to solve the general equation $ax + b = c$, of which the equation in the preceding example is a specific instance.

Each of the equations in these two examples can be solved easily by inspection. Nevertheless, these theorems are involved in determining these solutions. The fact is that we generally ignore this and, for example, go directly from the equation $x - 3 = 5$ to the equation $x = 5 + 3$ by ignoring the intermediate step $(x - 3) + 3 = 5 + 3$.

Example 4.   Solve the inequality $x - 5 < 8$. That is, find all whole number values of x for which this inequality is a true statement.

Solution: Applying Theorem A, add 5 to both sides of the given inequality to get the inequality

$$(x - 5) + 5 < 8 + 5.$$

Now use Theorem D to reduce the left side of this inequality to x and addition to reduce the right side to 13. We then have

$$x < 13.$$

Clearly, then, the solutions of the given inequality are the whole numbers 0, 1, 2, 3, 4, 5, 6, 7, 8, 9, 10, 11, and 12. This set of whole numbers is called the **solution set** of the given inequality.

Exercises 3.3

1. Explain why the symbols $3 \div 0$, $3 \div 2$, $2 - 3$, and $0 \div 0$ are all meaningless in the system of whole numbers. Explain why the symbol $0 \div 3$ has meaning and state its meaning.
2. Subtraction and multiplication share a (right) distributive property: $(x - y)z = xz - yz$. Use this distributive property to explain why $14x - 3x = 11x$. Do these operations share a left distributive property?
3. Which of the following statements do you think are true? (Assume that all the quotients are meaningful in the system of whole numbers.)
   (a) $a \div (b + c) = (a \div b) + (a \div c)$.
   (b) $(a + b) \div c = (a \div c) + (b \div c)$.
   (c) $a \div (b - c) = (a \div b) - (a \div c)$.
   (d) $(a - b) \div c = (a \div c) - (b \div c)$.
   (e) $a \div (b \cdot c) = (a \div b) \cdot (a \div c)$.
   (f) $a - (b \div c) = (a - b) \div (a - c)$.
   Each of these statements resembles a distributive property, but only some of them are true. Does it make any sense at all to say that multiplication is ambidextrous but that division is only right-handed?

4. Solve the following equations and indicate where and how in each solution you use any of Theorems C through F.
   (a) $x + 6 = 13$.   (b) $x - 4 = 7$.
   (c) $3x - 5 = 10$.   (d) $5x + 6 = 41$.

5. Theorem A has an analog for subtraction: *If x, y, and z are whole numbers, if x < y, and if z ≤ x, then x − z < y − z.* Theorem B has this analog for division: *If x, y, and z are whole numbers, if x < y, and both x ÷ z and y ÷ z are meaningful, then x ÷ z < y ÷ z.* Show how these theorems can be used to solve the following inequalities.
   (a) $x + 2 < 7$.   (b) $2x + 2 < 6$.

6. To prove Theorem E, we can begin by writing down the equation

$$x \cdot y = y \cdot x,$$

which we know is true because of what property? Since $y \neq 0$ (by hypothesis), what definition can we use to rewrite this equation in the following form?

$$(x \cdot y) \div y = x.$$

This is exactly what we wanted to prove, and so the proof is complete.

7. To prove Theorem F, we can begin by observing that, according to hypothesis, $x \div y$ is meaningful and so we can write

$$x \div y = x \div y$$

because of what property? Next, by applying what property do we see that this equation can be rewritten in the following form?

$$x = y \cdot (x \div y).$$

Then, by using what property can we write the following equation?

$$x = (x \div y) \cdot y.$$

Last, we use what property to tell us that

$$(x \div y) \cdot y = x?$$

3.4 Selected Consequences, III

The number zero has a murky history, and we know very little about the development of this important number idea. We do know that zero was regarded as a definite number by certain Hindu mathematicians as early as 900, but this may have been an isolated awareness. The idea that zero is as definite a number as the other whole numbers appears and disappears over the centuries. We read from a Hindu manuscript dated about 850 that a number "remains unchanged when divided by zero. . . ." This is obvious nonsense. As late as 1150 we read (again in a Hindu manuscript) that $3 \div 0$ represents an "infinite quantity"—more nonsense. And, of course, we read nothing much at all about zero in European works until even later. In this section we shall discuss one of the important features of this number in the system of whole numbers.

We know already that zero is an annihilator for multiplication—that zero times any number is zero. Let us phrase this statement in the form of a theorem.

Theorem G.
If one (or possibly both) of x and y is zero, then $xy = 0$.

It is a very important property of the system of whole numbers that the converse of Theorem G is also true. That is, we have a companion theorem to Theorem G.

Theorem H.
If x and y are whole numbers and if $xy = 0$, then at least one (and possibly both) of x and y is zero.

Proof of Theorem H:
Suppose that x and y are whole numbers and that $xy = 0$. Keep in mind that we are trying to prove that at least one of the numbers x and y is zero. Obviously, x, being a whole number, is either zero or it is not. We can therefore separate the proof into two cases.

Case 1:
x is equal to zero.

Case 2:
x is not equal to zero.

In Case 1 the theorem is obviously true, because if x is zero, then surely at least one of x and y is zero. Hence we only need to establish the truth of the theorem in Case 2, where x is not zero. So let us suppose that

$$xy = 0 \quad \text{and} \quad x \neq 0. \tag{1}$$

We already know (Theorem G) that

$$0 = x0. \tag{2}$$

We next combine equations (1) and (2) and get the equation

$$xy = x0. \tag{3}$$

Finally, we apply the cancellation property of multiplication to equation (3) and cancel the x. (Recall that $x \neq 0$ and we can cancel any nonzero number with respect to multiplication.) Hence we get

$$y = 0.$$

We have proved that if x is not equal to zero, then y is equal to zero. This means that *at least* one of x and y is zero. Thus we have proved the theorem.

In the system of whole numbers, if the product of two numbers is zero, then one or the other of the two numbers must equal zero. Contrast this with the situation in the arithmetic modulo six discussed in the preceding chapter. In that mathematical system it is possible to find two nonzero numbers whose product is 0.

Theorem H has applications to solving equations, as the following example shows.

Example 1. Solve the equation $(x - 1)(x - 2) = 0$.

Solution: The given equation tells us that the product of $x - 1$ and $x - 2$ is 0. Therefore, by Theorem H, either $x - 1$ must equal 0 or $x - 2$ must equal 0. If $x - 1 = 0$, then $x = 1$. If $x - 2 = 0$, then $x = 2$. Each of the numbers 1 and 2 is a solution.

Exercises 3.4 ——————————————————————

1. Use Theorem H to solve the following equations.
   (a) $(x + 2)(x - 2) = 0$.   (b) $(x - 1)(x - 2)(x - 3) = 0$.
2. Examine the argument based upon this sequence of statements.
   (1) Let $a = b \neq 0$.
   (2) Then $aa = bb$.
   (3) So $aa = ab$ (since $a = b$).
   (4) So $aa - bb = ab - bb$.
   (5) So $(a + b)(a - b) = (a - b)(b)$. [You can verify for yourself that $(a + b)(a - b) = aa - bb$.]
   (6) So $a + b = b$.

(7) So $b + b = b$. (Since $a = b$.)

(8) So $2b = b$ or $2b = 1b$.

(9) So $2 = 1$.

What is wrong?

3. In Exercise 7, Section 2.7, we asked you to construct the 5-multiplication table for arithmetic modulo five. Examine this table and see if it is possible to find nonzero objects x and y in that system such that $x \otimes y = 0$. Is it possible to find such nonzero objects in arithmetic modulo four?

4. In a hopeless but well-meant attempt to save the children of the Okefenokee the tribulations of the new math, Howland Owl invented the *aftermath*. The aftermath is a mathematical system containing as its only object the number 0. In this system subtraction and even division are binary operations. Also, it is impossible to find nonzero objects whose product is equal to 0. Explain all this.

3.5 Exponents

Before we continue with our discussion of the system of whole numbers, we must digress for a moment to discuss an important method of symbol abbreviation. Very often, in the study of whole numbers, we must deal with products like $2 \cdot 2 \cdot 2 \cdot 2$ and $3 \cdot 3 \cdot 3 \cdot 3 \cdot 3$. Because such products are somewhat awkward to write down, we abbreviate them by writing 2^4 in place of $2 \cdot 2 \cdot 2 \cdot 2$ and 3^5 in place of $3 \cdot 3 \cdot 3 \cdot 3 \cdot 3$. In the symbol 2^4, the number 4 is called the **exponent** or power and is said to occupy the superscript position. The number 2 in the symbol 2^4 is called the **base.**

The important thing to remember is that the exponential symbol n^m is nothing more than an abbreviation for a product, and when this symbol is seen it is the product itself that one should think of. For example, when you see the symbol "$\$$," you think of the word this symbol abbreviates. The symbol is simply the key that brings the word "dollars" to your mind. In exactly the same way the symbol 4^3 should bring the product $4 \cdot 4 \cdot 4$ to mind. We rarely think in terms of the abbreviations themselves. We only use the abbreviations to tell us what to think about. Students who have difficulty with exponents are most often trying to think in terms of the abbreviations themselves instead of in terms of what the abbreviations abbreviate.

It should be clear now what the symbol x^y means when both x and y are nonzero whole numbers. If x is zero and y is nonzero, then x^y represents the product $0 \cdot 0 \cdot 0 \cdot 0 \cdots 0$ (y factors) and so represents 0. But if y is zero and x is nonzero, then x^y cannot properly be thought of as an abbreviation of a product. By common agreement (that is, by definition) we assign the meaning 1 to all symbols of the form x^0, where x is nonzero. (We shall explain why we do this in a moment.)

Finally, we shall give the symbol 0^0 no meaning at all. For us, 0^0 will be a nonsense symbol.

The exponential symbols (except for x^0) are nothing more than abbreviations for longer products, but there are many different ways that such products could be abbreviated. Why was this particular method chosen over all alternatives? The answer to this lies in the fact that these exponential symbols lend themselves to symbol manipulation. For example, consider the product of 2^3 and 2^4. We write

$$
\begin{aligned}
2^3 \cdot 2^4 &= (2 \cdot 2 \cdot 2)(2 \cdot 2 \cdot 2 \cdot 2) \\
&= 2 \cdot 2 \cdot 2 \cdot 2 \cdot 2 \cdot 2 \cdot 2 \\
&= 2^7 \\
&= 2^{3+4}.
\end{aligned}
$$

This example illustrates the general fact that

$$
x^p \cdot x^q = x^{p+q}.
$$

Thus, to multiply powers of the same base, all you have to do is add the exponents.

Next, suppose the bases are different but the powers are the same. For example, $3^2 \cdot 7^2 = ?$ If we replace the exponential symbols by the products they abbreviate, then we can rearrange the factors and simplify the product:

$$
\begin{aligned}
3^2 \cdot 7^2 &= (3 \cdot 3)(7 \cdot 7) \\
&= (3 \cdot 7)(3 \cdot 7) \\
&= (3 \cdot 7)^2.
\end{aligned}
$$

This example suggests that to multiply the same powers of different bases, multiply the bases, and raise this product to the common power:

$$
x^p \cdot y^p = (x \cdot y)^p.
$$

The last theorem concerning these exponential symbols that we shall need has to do with division. Suppose that we want to divide different powers of like bases. For example, consider $3^5 \div 3^2 = ?$ Again, replace the exponential symbols by the products they abbreviate and rearrange:

$$
\begin{aligned}
3^5 \div 3^2 &= (3 \cdot 3 \cdot 3 \cdot 3 \cdot 3) \div (3 \cdot 3) \\
&= 3 \cdot 3 \cdot 3 \\
&= 3^3 \\
&= 3^{5-2}.
\end{aligned}
$$

This example illustrates the fact that to divide different powers of the same base, simply subtract exponents—if it is possible to do so. That is,

$$
x^p \div x^q = x^{p-q}, \qquad \text{if } p \geq q.
$$

Suppose that $x \neq 0$ and look again at the equation

$$x^p \cdot x^q = x^{p+q}.$$

We can see from this equation why we were induced to define x^0 to mean 1. Suppose that in this equation q is 0. Then we would have

$$x^p \cdot x^0 = x^{p+0}.$$

But x^{p+0} is equal to x^p. So, what we have is

$$x^p \cdot x^0 = x^p.$$

Since the only way this equation can be true is for x^0 to be 1, we define it to be 1.

In the rest of this chapter, we shall be dealing with ideas that come under the heading of *elementary number theory*. Our treatment of elementary number theory will be considerably more informal than our treatment in this and the two preceding sections of the elementary consequences of the basic properties.

Exercises 3.5

1. Rewrite each of these products using exponential notation.
   (a) $3 \cdot 3 \cdot 2 \cdot 3 \cdot 2$.   (b) $2 \cdot 3 \cdot 5 \cdot 2 \cdot 3 \cdot 5 \cdot 2$.
   (c) $2 \cdot 3 \cdot 5^2 \cdot 7^2 \cdot 2 \cdot 3 \cdot 5 \cdot 7$.
2. Use the theorems to rewrite each of these products in a simpler exponential form. For example, $2^2 \cdot 3^2 \cdot 2^4 = 2^6 \cdot 3^2$.
   (a) $2^4 \cdot 2^0 \cdot 3^2 \cdot 2^2 \cdot 5^4 \cdot 3^3$.   (b) $2^0 \cdot (313^2 \cdot 55^5)^0$.
   (c) $(2^2 \cdot 2^4) \div 2^3$.
3. You have seen the famous equation $E = mc^2$ many times before. Does this equation say that $E = mcc$ or that $E = mcmc$? How should you insert parentheses if you wanted to make this equation mean what it does not mean?
4. Suppose that x, p, and q are all nonzero whole numbers. If necessary, experiment with a few specific numbers in order to complete the statement $(x^p)^q = x^?$.
5. How many parents does a person have? How many grandparents? How many great grandparents? How many great-great-grandparents? How many great-great-great-grandparents? If you assume three generations every 100 years, how many ancestors do you have going back 200 years?
6. Each number of the form 1 followed by 0's can be written as a product of 10's and can then be abbreviated using exponential notation. Write each of the numbers 10, 100, 1000, 10,000, and 100,000 as a power of 10. The number 10^{56} is a number of the form 1 followed by 0's; how many 0's? The number 1 has the form 1 followed by a certain number of 0's. How is 1 written as a power of 10?

7. A *googol* is defined to be 1 followed by 100 zeros. Express a googol as a power of 10. A *googolplex* is 1 followed by a googol of zeros. Express a googolplex as a power of 10. A googol is a very large number, for it has been estimated that the total number of electrons in the universe is only about 10^{79}.

3.6 Prime Numbers

There are a great many ways of classifying whole numbers. You know, for example, of the classification of all whole numbers into even and odd numbers. Another classification is into prime and composite numbers. The study of prime numbers is ancient and still occupies a central part of number theory. In this section we shall discuss these numbers.

One of the most important relations on whole numbers is the one called *is a divisor of*. When we say that x is a divisor of y, we mean that

$y = x \cdot$ (some other whole number).

Thus 4 is a divisor of 20 because $20 = 4 \cdot 5$. But 4 is not a divisor of 21 because there is no whole number x such that $21 = 4 \cdot x$.

There are other ways to express the fact that a number x is a divisor of a number y. We can say that *y is a multiple of x*. For example, 20 is a multiple of 4 and 18 is a multiple of 9. We can also say that *y is divisible by x*. Thus 20 is divisible by 4 and 18 is divisible by 9. The statements "x is a divisor of y," "y is a multiple of x," and "y is divisible by x" all mean exactly the same thing.

One of the useful things to know about a whole number is the divisors of that number. For example, the divisors of 20 are 20 itself, 1 (which is a divisor of every number), 2, 4, 5, and 10. In fact, there are methods of classification of whole numbers on the basis of the divisors of the numbers. Of these the most important is the classification of whole numbers with regard to whether or not they are **prime numbers.** A prime number is a number that has no divisors other than the number itself and the omnipresent divisor 1. In Table 3.1 we have listed all the prime numbers less than 1000. Incidentally, although it might seem reasonable to call 1 a prime, if we were to do so there would be many times when we would have to make statements like, "For all prime numbers except 1, it is true that . . . ." To avoid having to make this exception, we do not call 1 a prime.

A whole number greater than 1 that is not a prime number is called a **composite** number. Every number greater than 1 is either composite or prime, and 0 and 1 are neither composite nor prime.

There are many interesting questions concerning prime numbers. For example,

Table 3.1 Table of Primes Less Than 1000

2	101	211	307	401	503	601	701	809	907
3	103	223	311	409	509	607	709	811	911
5	107	227	313	419	521	613	719	821	919
7	109	229	317	421	523	617	727	823	929
11	113	233	331	431	541	619	733	827	937
13	127	239	337	433	547	631	739	829	941
17	131	241	347	439	557	641	743	839	947
19	137	251	349	443	563	643	751	853	953
23	139	257	353	449	569	647	757	857	967
29	149	263	359	457	571	653	761	859	971
31	151	269	367	461	577	659	769	863	977
37	157	271	373	463	587	661	773	877	983
41	163	277	379	467	593	673	787	881	991
43	167	281	383	479	599	677	797	883	997
47	173	283	389	487		683		887	
53	179	293	397	491		691			
59	181			499					
61	191								
67	193								
71	197								
73	199								
79									
83									
89									
97									

Euclid proved that there are infinitely many prime numbers, but it has not been possible to actually determine more than a finite number of primes. The problem is that it is extremely difficult to determine divisors of very large numbers. Generally speaking, this requires prodigious amounts of computation. Computers have been applied to this task and have met with some success. For example, the largest known prime number is the number $2^{19937} - 1$. This number was proved to be a prime by using a computer. Because the number has 6002 digits, it is unlikely that the fact that this number is a prime could have been confirmed by computational means without the aid of the computer.

The flow chart in Figure 3.2 gives a procedure that can in theory be used to determine whether or not an odd number N is a prime. (We don't have to bother with even numbers because the only even prime is 2.) However, the procedure given there is too crude for actual use. Much of the work in finding whether a very large number N is a prime goes into finding more efficient procedures.

Another question about prime numbers that has enjoyed continued attention over the years is the famous **Goldbach Conjecture.**[1] Goldbach's conjecture was

---

[1] In 1742, C. Goldbach (1690–1774) posed this conjecture to the great eighteenth-century mathematician, Leonhard Euler. He also discovered theorems in geometry that express the principles upon which the Wankel engine is based.

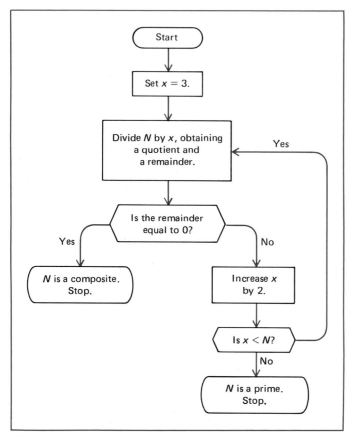

Figure 3.2 *Flow chart for determining if an odd number N greater than 3 is a prime.*

that every even whole number greater than 2 can be written as the sum of two primes.[2] For instance,

$$
\begin{array}{lll}
4 = 2 + 2 & 12 = 7 + 5 & 20 = 13 + 7 \\
6 = 3 + 3 & 14 = 7 + 7 & 22 = 11 + 11 \\
8 = 5 + 3 & 16 = 11 + 5 & 24 = 11 + 13 \\
10 = 5 + 5 & 18 = 11 + 7 & 26 = 13 + 13.
\end{array}
$$

This conjecture has been shown to be true for all even numbers small enough to be amenable to computational techniques, and thus one might guess that it is true in general. The conjecture is very famous and mathematicians have tried without success to prove that it is true. It is interesting that in trying to prove

[2]The conjecture is obviously false for odd numbers. Consider 11 or 17.

that every even number greater than two could be written as the sum of two primes, a Russian mathematician named Schnirelmann (1905–1938) proved in 1931 that every even whole number greater than two can be written as the sum of not more than 300,000 primes! A contemporary Russian, Vinogradoff, made another advance by proving that there is a whole number N (the value of N is not known, it is only known that such a number exists) such that every even whole number greater than N can be written as the sum of at most four primes. This is the present state of the problem.

Another famous and long unsolved conjecture has to do with what are called *twin primes*. Two whole numbers x and y are called **twin primes** if they are consecutive odd primes. For example, the following pairs are twin primes:

3 and 5
5 and 7
11 and 13
17 and 19
29 and 31.

The conjecture is that there are infinitely many pairs of twin primes. Whether there exist infinitely many such pairs of primes is unknown, and the question has not been resolved, even though it has received a great deal of attention over the years.

These examples illustrate that one need not delve very deeply into the theory of numbers in order to come up with a question that cannot, at present, be answered. Many unanswered questions lie very near to the surface.

Exercises 3.6

1. By inspection find all the divisors of the numbers 12, 145, and 144.
2. To say that x is an even prime says what about x?
3. Using the table of primes, find all pairs of twin primes between 1 and 1000.
4. Verify Goldbach's Conjecture for the even whole numbers 38, 40, 42, 68, 70, 72, 146, 148, 150, 638, 640, and 642.
5. It can be proved that every odd whole number greater than 5 can be written as the sum of three primes. Verify this for the odd whole numbers from 5 to 25. Also verify that 143 and 301 are sums of three primes.
6. Consider the relation defined on the set of whole numbers and which is called *is a divisor of*.
   (a) Find all whole numbers that are related to 16 by this relation.
   (b) Find all whole numbers to which 16 is related by the relation.
   (c) Are any whole numbers related to themselves by this relation?
   (d) Is the relation transitive?
   (e) Show that the relation is not symmetric by giving an example.

7. Consider the relation *is a multiple of* defined on the set of whole numbers.
   (a) To which whole numbers is 16 related?
   (b) Which whole numbers are related to 16?
   (c) Is this an equivalence relation?
*8. As you move out in the list of whole numbers, the primes get rarer and rarer in the sense that, given any whole number n, by going out far enough in the list you can find a succession of n composite numbers. For example, there is a place in the list where there are 50 composite number numbers in a row. Explain why each of the following numbers is composite:

$$51! + 2, \ 51! + 3, \ 51! + 4, \ldots, \ 51! + 51$$

and explain why there are 50 successive numbers in this list. (The symbol 51! is an abbreviation for the product of all the whole numbers from 1 to 51; for example, $4! = 4 \cdot 3 \cdot 2 \cdot 1$.) How would you find 100 successive composite numbers?

9. Prove that after you pass by 2 and 3 in the list of whole numbers you can never find two successive prime numbers.

10. There have been many attempts to find formulas that would generate prime numbers. Here are some examples of the more obvious failures. Find out when these formulas fail to produce prime numbers.
   (a) $2^n + 1$, where n is a nonzero whole number.
   (b) $2^n - 1$, where n is a nonzero whole number.
   (c) $n^2 - n + 11$, where n is a nonzero whole number.
   (d) $n^2 - n + 41$, where n is a nonzero whole number.

11. The flow chart in Figure 3.2 is very inefficient. The decision step "Is $x < N$?" that terminates the process should be replaced by "Is $x < \sqrt{N}$?" for greater efficiency. For example, in testing 145 it is not necessary to check divisibility by all odd numbers up to 143, it is only necessary to check odd numbers up to 13. Explain why this is so.

3.7 Prime Factorizations

The prime numbers are the basic building blocks from which all whole numbers (greater than one) are built. In this section we shall discuss this idea.

The prime numbers are basic in the sense that each whole number greater than 1 can be expressed as a product of prime numbers. For example,

$$20 = 2 \cdot 2 \cdot 5 \qquad 198 = 2 \cdot 3 \cdot 3 \cdot 11$$
$$45 = 3 \cdot 3 \cdot 5 \qquad 144 = 2 \cdot 2 \cdot 2 \cdot 2 \cdot 3 \cdot 3.$$

The prime numbers can be regarded as the building blocks from which other whole numbers greater than 1 can be obtained by using multiplication. This intuitively obvious fact is a part of the statement of an important theorem called the "Fundamental Theorem of Arithmetic."

The Fundamental Theorem of Arithmetic.

Every whole number greater than 1 can be written as a product of prime numbers. Apart from the order in which these prime number factors are written down, there is only one such way to write each whole number as a product of prime numbers.

Example 1. The number 30 can be expressed as $2 \cdot 3 \cdot 5$ and as $3 \cdot 2 \cdot 5$, but these are really the same way of writing 30 as a product of primes; only the order of writing down the prime factors is different.

When we write a whole number as a product of primes, we call this product the **prime factorization** of the whole number. Thus $2 \cdot 3 \cdot 5$ is the prime factorization of 30 and $2 \cdot 2 \cdot 5$ or $2^2 \cdot 5$ is the prime factorization of 20.

The Fundamental Theorem implies that every whole number greater than 1 is divisible by at least one prime. If such a number is a prime, then it is divisible by itself. If the number is composite, then according to the Fundamental Theorem that number can be written as a product of primes and so is divisible by each of its prime factors. We shall use this simple fact to prove a theorem mentioned in the last section.

Theorem.

There are infinitely many prime numbers.

Proof:

We shall use the *indirect method* to prove this theorem. The proof we shall use is a classical example of the elegant simplicity of many indirect proofs, and is attributed to Euclid. We shall begin by making the tentative assumption that there are only a finite number of primes. Then, on the basis of this tentative assumption, we shall draw logically valid conclusions, one of which will be a contradiction to something we already know. This contradiction will mean that the tentative assumption must have been false, and this will mean that there are infinitely many prime numbers. We shall present this proof as a sequence of numbered statements.

1. There are only finitely many primes. (This is the tentative assumption on the basis of which we shall reach a contradiction.)
2. There is a largest prime, say, p. (Every finite set of whole numbers contains a largest number.)
3. Let n be the product of the finitely many primes: $n = 2 \cdot 3 \cdot 5 \cdot 7 \cdot 11 \cdots p$. (This is simply the definition of the symbol n and does not require justification.)

4. Let $m = n + 1$. (This is simply the definition of the symbol m and does not require justification.)
5. m is divisible by some prime, say q. (By the remark preceding this theorem, every whole number greater than 1—and m is greater than 1—is divisible by some prime number.)
6. $m = qx$, where x is some whole number. (This is simply another way of expressing the fact that m is divisible by q.)
7. $n = qy$, where y is some whole number. (The number n is defined to be the product of all primes. q is a prime and so is a divisor of n.)
8. $qx = qy + 1$. [From statements (4), (6), and (7).]
9. $qx - qy = 1$. [This follows from statement (8) by using the definition of subtraction.]
10. $q(x - y) = 1$. [This follows from statement (9), since subtraction and multiplication share a distributive property.]

But now we have reached a contradiction. Statement 10 states that the prime number q divides 1, and this is false. As the argument that leads from tentative assumption to this contraction is valid, it can only be that the tentative assumption is incorrect. Thus it is incorrect that there are only finitely many primes; therefore, what must be true is that there are infinitely many primes. This is precisely the statement that we were to prove and the proof is complete.

Exercises 3.7

1. Write out the prime factorization of each whole number from 50 to 65.
2. Why does the fact that we can write 24 as $2^3 \cdot 3$ and as $2^3 \cdot 3 \cdot 5^0$ not contradict the Fundamental Theorem?
3. How many divisors does 5^{57} have?
4. Duplicate the proof in the text to show that it is false that the only primes are the numbers 2, 3, 5, 7, 11, and 13.
5. How can you tell whether or not a whole number is a perfect square simply by looking at its prime factorization?
*6. A whole number greater than one is called *square-free* if it has no divisors that are perfect squares. How can you tell whether or not a number is square-free simply by looking at its prime factorization? Write out all the square-free whole numbers between 1 and 30.
*7. There are some divisibility tests that can be used to help locate small divisors of numbers. Among these are the following:

A number is divisible by 2 (or by 5) if its last digit is divisible by 2 (or by 5).
A number is divisible by 4 (or by 10) if the number formed from its last two digits is divisible by 4 (or by 10).

A number is divisible by 8 if the number formed by its last three digits is divisible by 8.

A number is divisible by 3 (or by 9) if the sum of its digits is divisible by 3 (or by 9).

Use these tests to determine divisors of the following numbers.

(a) 73,624.   (b) 123,456,789.

(c) 2468.   (d) 12,618.

(e) 354,629,805,041,020.   (f) 983.

*8. Use the divisibility tests of the preceding exercise to find the prime factorizations of the following numbers:

(a) 1080.   (b) 1620.   (c) 2754.   (d) 2244.

*9. In Chapter 0 we gave a flow chart for determining whether a whole number is divisible by 3. Construct such charts for the other divisibility tests given in Exercise 7.

3.8   Greatest Common Divisor and Least Common Multiple

In this section we shall use prime factorizations to introduce some number-theoretic ideas that will be useful in our study of rational numbers in Chapter 6.

Suppose that x and y are two whole numbers, neither of which is zero. The greatest number that is a divisor of both x and y is called the **greatest common divisor** of x and y. For instance, the greatest number that divides both 20 and 45 is 5. We can see why 5 is the greatest common divisor of these numbers by writing 20 and 45 in their prime factorizations and then comparing these prime factorizations:

$$20 = 2^2 \cdot 5 \quad \text{and} \quad 45 = 3^2 \cdot 5.$$

We see that the only prime numbers that divide 20 are 2 and 5, whereas the only prime numbers that divide 45 are 3 and 5. Hence the only prime number that will divide *both* 20 and 45 is 5. Then we observe that the highest power of 5 that will divide both numbers is the first power. We conclude, therefore, that 5^1 or 5 is the greatest common divisor of 20 and 45. Here are some more examples of finding greatest common divisors using prime factorizations. From now on we shall abbreviate greatest common divisor by GCD.

Example 1.   Find the GCD of 45 and 81.

Solution: We write $45 = 3^2 \cdot 5$ and $81 = 3^4$. Then we observe that the only prime that divides both numbers is 3. Also, the largest power of 3 that divides both numbers is 3^2. Hence 3^2 or 9 is the GCD of these numbers.

Example 2.   Find the GCD of $2 \cdot 3^2 \cdot 5$ and $2^3 \cdot 3^3 \cdot 7$.

Solution: The only primes that divide both numbers are 2 and 3. The greatest power of 2 that divides both numbers is the first power. The greatest power of 3 that divides both numbers is the second power. Hence the GCD is the product of 2^1 and $3^2 : 18$.

The concept of greatest common divisor has a spin-off concept of importance. If x and y are whole numbers whose greatest common divisor is 1, then we say that x and y are **relatively prime.** Thus 15 and 14 are relatively prime whole numbers, since their GCD is 1.

If x and y are whole numbers, then the smallest whole number that is divisible by both x and y is called the **least common multiple** of x and y. The least common multiple of 4 and 6 is 12, since 12 is divisible by both 4 and 6 and is the least number with this property. The least common multiple of 12 and 18 is 36, since 36 is divisible by both 12 and 18 and no number less than 36 is divisible by both 12 and 18.

Prime factorizations can also be used to find least common multiples. (We shall henceforth abbreviate least common multiple by LCM.) For example, consider the numbers 12 and 15. Since

$$12 = 2^2 \cdot 3 \quad \text{and} \quad 15 = 3 \cdot 5,$$

we see that any number which is divisible by *both* 12 and 15 must involve all of the primes 2, 3, and 5 in its prime factorization. Hence the LCM of 12 and 15 has the general form $2^a \cdot 3^b \cdot 5^c$. Moreover, the least number of this form that is divisible by both $2^2 \cdot 3$ and $3 \cdot 5$ is $2^2 \cdot 3^1 \cdot 5^1$; thus the LCM of 12 and 15 is 60.

Example 3.   The LCM of 45 and 81 is found by first writing these numbers in their prime factorizations: $45 = 3^2 \cdot 5$ and $81 = 3^4$. Any number that is a multiple of both 45 and 81 must, therefore, have the general form $3^a \cdot 5^b$. Of all such numbers the LCM is the one with the smallest possible exponents. The least a can be is 4 (or else this number would not be divisible by 3^4), and the least b can be is 1. Hence the LCM of 45 and 81 is $3^4 \cdot 5^1$ or 405.

Example 4.   Find the LCM of $2 \cdot 3^2 \cdot 5^3$ and $2^3 \cdot 3^3 \cdot 7$.

Solution: The LCM must involve the primes 2, 3, 5, and 7 in order to be divisible by both of the given numbers. So, the LCM has the form $2^a \cdot 3^b \cdot 5^c \cdot 7^d$. Now select a, b, c, and d to be as small as possible so that the LCM will be divisible by both of the given numbers. The exponents desired are $a = 3$, $b = 3$, $c = 3$, and $d = 1$. The LCM is $2^3 \cdot 3^3 \cdot 5^3 \cdot 7^1$.

Exercises 3.8

1. Compute the GCD and LCM of each pair of numbers by using prime factorizations.
   (a) 12 and 30. (b) 70 and 462.
   (c) 198 and 144. (d) 1350 and 120.

2. Use your answers to Exercise 1 to demonstrate the truth of the following theorem.

   Theorem.
   If x and y are whole numbers greater than zero, then $(GCD)(LCM) = x \cdot y$.

3. If x and y are relatively prime, then what is their LCM?

4. Find three numbers between 100 and 200 that are relatively prime to 144.

5. Suppose a and b are relatively prime and that c and d are relatively prime.
   (a) Are ac and bd relatively prime?
   (b) Are $a + c$ and $b + d$ relatively prime?
   (c) Are a^2 and b^2 relatively prime?
   (d) Are a and $a + b$ relatively prime?
   (It is anticipated that you will answer these questions experimentally rather than by proving them to be true or false.)

6. Let * denote the operation defined on the nonzero whole numbers according to the rule

   $$x * y = GCD(x, y).$$

   Answer the following questions about this operation.
   (a) Does this process always result in exactly one number no matter which nonzero whole numbers x and y represent? That is, is * really a binary operation?
   (b) Is * commutative? Is it associative?
   (c) Is there an identity for this operation? That is, is there a whole number i such that $x * i = i * x = x$ no matter which whole number x represents?
   (d) Does the cancellation property hold for this operation? That is, if $x * z = y * z$, can you conclude that $x = y$?

7. Replace the "GCD operation" in Exercise 6 by the operation * defined by the rule:

   $$x * y = LCM(x, y)$$

   and answer the same questions for this operation.

8. Suppose n, x, and y are whole numbers such that x divides n and y divides n. Then does it necessarily follow that the product of x and y divides n? Find some examples to substantiate your answer. By using the ideas of this section, it is possible to impose a condition involving x and y so that it will necessarily follow that n will be divisible by xy. Can you find such a condition?

9. Construct a flow chart for finding the greatest common divisor of two nonzero whole numbers.
10. Construct a flow chart for finding the least common multiple of two nonzero whole numbers.

3.9 Fermat's Last Problem

No study of number theory would be complete without mention of the father of modern number theory and one of his most famous problems. Pierre de Fermat (French, 1601?–1665) worked in many fields of mathematics and we shall run across his name again, but he is most famous for his work in number theory. He is responsible for the most famous unresolved problem in all of mathematics. In this section we shall discuss this problem, which is called "Fermat's Last Problem."

Consider the right triangle shown in Figure 3.3. The famous theorem from geometry called the Pythagorean theorem states that the lengths of the sides of such a right triangle are related by the equation

$$a^2 + b^2 = c^2,$$

where c denotes the length of the side of the triangle opposite the right angle and a and b represent the lengths of the sides adjacent to the right angle. A triple of numbers (a, b, c) is called a **Pythagorean triple** if the numbers a, b, and c satisfy such an equation.

Example 1. The triple $(3, 4, 5)$ is a Pythagorean triple, since $3^2 + 4^2 = 5^2$. Other Pythagorean triples are $(5, 12, 13)$, $(7, 24, 25)$, and $(51, 140, 149)$.

If we now generalize the problem of finding Pythagorean triples, we can ask the following question: If n is a whole number greater than 2, is it possible to find nonzero integers[3] x, y, and z such that $x^n + y^n = z^n$? Fermat (so the story

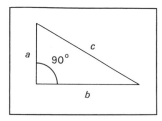

Figure 3.3

[3]An integer is a number that is either a whole number or the negative of a whole number: -1, -2, -3, -4, ... We shall discuss integers in Chapter 6.

goes) wrote in the margin of one of his mathematics books that he had found an elegant proof of the fact that it is impossible to find such numbers x, y, and z if n is greater than 2. Unfortunately, there was not enough room in the margin to contain the proof itself. Fermat died before he could write out his proof—hence the name, Fermat's Last Problem.

The important point here is that no one has ever been able to prove that there are integers x, y, and z such that $x^n + y^n = z^n$, when $n > 2$, or that there are not any such triples of integers. It is most probable that Fermat was mistaken and that his "elegant proof" was somehow in error. This problem has received an incredible amount of attention, and very little real progress has been made toward answering it. It has been shown, however, that no integers x, y, and z exist such that $x^n + y^n = z^n$ if n is between 2 and 250,000. This result was achieved by using a computer.

Exercises 3.9

1. Prove that if (a, b, c) is a Pythagorean triple, then so is $(2a, 2b, 2c)$. Prove in general that if (a, b, c) is a Pythagorean triple, then so is (na, nb, nc), where n is a nonzero whole number. This proves that there are infinitely many Pythagorean triples.

2. Let x and y be nonzero whole numbers with $x > y$. Prove that the triple $(x^2 - y^2, 2xy, x^2 + y^2)$ is a Pythagorean triple. Use this fact to find three Pythagorean triples.

3. Prove that not all three numbers of a Pythagorean triple can be odd. Can exactly two be even? Can all three be even?

*4. Another famous unresolved problem has to do with what are called "perfect numbers." A whole number is said to be **perfect** if it is equal to the sum of its proper divisors. For example, 6 is perfect, since the proper divisors of 6 are 1, 2, and 3, and $6 = 1 + 2 + 3$. Show that 28 and 496 are perfect numbers. Euclid proved that if $2^{n+1} - 1$ is a prime number, then $(2^{n+1} - 1)(2^n)$ is a perfect number. Use this theorem to prove that 8128 is a perfect number. That is, show that 8128 has this particular form. Whether or not there are any odd perfect numbers is unknown. It is not even known whether there are infinitely many even perfect numbers.

4

The Theory of Sets

From the time of the ancient Greeks the notion of infinity has posed the most serious kinds of difficulties to mathematicians and scientists. There appeared to be no logical way to handle the idea of an infinite set, and perhaps the principal reason for this was that the notion of an infinite set is entirely foreign to our experience and environment. Trying to gain an understanding of infinite sets by approaching them on the basis of the understanding we have for finite sets is impossible. The best mathematicians recognized this and, not knowing what to do, declared that mathematicians should not be allowed to become involved with infinite sets at all.

The trouble with this decision is that most of the important parts of modern mathematics would be impossible without infinite sets. By the beginning of the nineteenth century, mathematics was approaching a crisis point due, in large part, to the misuse and mistrust of infinite sets. Consequently, when, in 1874, Cantor presented his ingenious and original method of dealing with infinite sets in a mathematically rigorous way, the effect upon mathematics was staggering. After some initial resistance, his ideas were accepted readily and work was immediately begun to reestablish the notoriously weak foundations of mathematics upon this new and precise theory of sets.

However, it was not long before weak spots began to show up in this theory and, in fact, work is still going on to shore up these logical weaknesses. But set theory is firmly established in mathematics now and will remain so. It is an indispensable part of modern mathematics. In this chapter we shall discuss some of the more elementary aspects of Cantor's theory of sets.

4.1 The Concept of Set

When Georg Cantor (1845–1918) presented his theory of sets, he wrote that a set was "a bringing together into a whole of definite well-established objects of our perception or thought . . . ." Now, this can in no sense be taken as a careful definition of what we mean by a set of objects. This partial description does, however, hint at two important aspects of this concept. First, a set is a "bringing together into a whole." That is, a set is an entity in itself. Although a set may contain many different and very diverse objects, when we think of the set of those objects we are thinking of a single entity. Second, a set is a bringing together of "well-established objects" and this means that when we want to talk about a specific set, we must have some way of determining which objects belong to the set and which objects do not belong to the set. For example, we may talk about the set of all digits because this verbal description of this set enables us to determine which numbers are in the set and which are not.

If we do not use the set concept except in natural and ordinary situations, we may as well take the concept of set to be undefined; in such cases there is nothing to be gained by attempting to give a carefully drawn mathematical definition for the concept. In all of our work in this book, the concept will apply to very ordinary sets, and so we need not be too concerned over the exact meaning of the term *set*. However, if we try to use the concept in less natural instances, we can easily run into logical complications if we do not somehow restrict the meaning of the concept. In the next section we shall demonstrate the necessity for sometimes being very careful about how we use the term. Except for Section 4.2, however, we may use the term *set* freely, without worrying about becoming enmeshed in logical contradictions.

We most often symbolize sets by capital Latin letters. Thus we may talk about the set W of all whole numbers, the set E of all even whole numbers, and the set D of digits. Objects belonging to sets are symbolized by small Latin letters. Thus we might talk about an object s belonging to a set S. To symbolize the fact that the object s belongs to the set S, we write

$$s \in S.$$

The symbol $\in$ is read "is an object of" or "belongs to." This is an invented symbol, but it derives from the Greek letter ε (epsilon), which is the first letter of the Greek word for "is."

Example 1. If D represents the set of digits, then $0 \in D$, $1 \in D$, . . . , $8 \in D$, and $9 \in D$. But $10 \notin D$ and $11 \notin D$.

We can describe sets using the brackets { and } as illustrated in these examples.

Example 2.   The set of all digits can be written as {0, 1, 2, 3, 4, 5, 6, 7, 8, 9}.

Example 3.   The set of all whole numbers can be written as {0, 1, 2, 3, 4, 5, . . .}. In this case we cannot hope to list all of the objects of this set, so the best we can do is actually to list enough of these objects so that the reader can get the idea of which objects the set contains, and then use the ellipsis symbol (. . .) to indicate that the listing is to continue in the manner indicated by the objects that were actually listed. The set of all even whole numbers would be described as {0, 2, 4, 6, 8, . . .}.

Example 4.   The set of all whole numbers less than 100 contains too many objects to be listed conveniently, so write {0, 1, 2, 3, 4, . . . , 99} instead. The ellipsis symbol indicates that some of the objects have not been written down.

There is one set that is unusual enough to deserve special attention. This is the set that we call the **empty set** and that, as its name implies, does not contain any objects at all. For example, the empty set is the set of all whole numbers less than 0. It is the set of all people in the United States who are more than 250 years old. This set is usually denoted by the special symbol $\varnothing$.

Exercises 4.1 _____

1. Think of some words that are commonly used as synonyms for the term *set*. For example, a set of people who are related is sometimes called a *clan*.
2. Let A denote the set of cloth objects and B denote the set of white objects. Name an object x such that
   (a) $x \in A$ but $x \notin B$.   (b) $x \in A$ and $x \in B$.
   (c) $x \notin A$ and $x \notin B$.   (d) $x \notin A$ but $x \in B$.
3. Describe a set S that contains exactly four people. Then find two objects x and y such that $x \in S$ and $y \notin S$.
4. Use the bracket notation to describe the following sets.
   (a) The set of last words on pp. 10, 11, and 12 of this text.
   (b) The set of all whole numbers less than your age.
   (c) The set of all whole numbers greater than your age and less than 10,000,000.
5. Let S denote the set of whole numbers for which the inequality $x - 3 < 10$ is a true statement. Insert $\in$ or $\notin$ in the spaces as appropriate:
   (a) 3 [ ] S.   (b) 4 [ ] S.   (c) 5 [ ] S.
   (d) 12 [ ] S.   (e) 13 [ ] S.   (f) 14 [ ] S.
6. The empty set may be described as the set of all unicorns in the Foosland Zoo. Describe the empty set using the concept of
   (a) Whole number.   (b) Disjunction.   (c) Addition.
7. What is the distinction between the sets $\varnothing$ and $\{\varnothing\}$?

8. Use the bracket notation to describe the set of whole numbers that are solutions of the inequality
   (a) $x - 3 > 10$.   (b) $2 - x < 5$.   (c) $2x - 1 \leq 15$.
9. It is most often the case that all of the sets under discussion at any one time will consist of objects drawn from some all-encompassing set called the **universal set**. For example, throughout Chapter 3 the universal set was the set of all whole numbers. List the objects of the set S of all numbers that are less than 10 if the universal set is
   (a) The set of all even whole numbers.
   (b) The set of all whole numbers.
   (c) The set of all odd whole numbers.
   (d) The set of all whole numbers that are less than 5.
   This exercise demonstrates the necessity for always making clear just what the universal set is during any particular discussion involving sets.
*10. Use the bracket notation to describe the set of whole numbers for which the statement is true.
   (a) x is less than 5 and x is greater than 2.
   (b) If x is less than 5, then x^2 is also less than 5.
   (c) Either x is less than 5 or x is greater than 5.
   (d) It is false that x is less than 5.

$4.2^\dagger$ Some Paradoxes

In the last section we said that we were going to take the notion of a set as an undefined concept. We know that every deductive system must contain such undefined concepts, and, because it is quite difficult to find a good definition of set in terms of more primitive ideas, we may as well start with set as being undefined. For most elementary purposes in mathematics, and for everything we have to do with sets in this book, this is a perfectly acceptable thing to do. But in the larger context of set theory generally, the lack of definition does present serious logical complications. If we do not somehow place a restriction upon what can be called a set and what cannot be called a set, then we run into logical paradoxes. In this section, therefore, we shall introduce the notion of a paradox and then show how an immoderate application of the term *set* results in a paradox.

A **paradox** is a seemingly self-contradictory statement. The logic used in mathematics is a consistent logic in the sense that this logic does not permit the existence of a statement that is self-contradictory. Hence, whenever a paradox is found, it is necessary to explain how the paradoxical statement is only *seemingly* self-contradictory. In this section we are going to present a short list of famous paradoxes. However, we shall not explain them all because this involves a more sophisticated study of mathematical logic, and we are not prepared to enter into these studies. We can only state that all known paradoxes have been explained to the satisfaction of most mathematical logicians.

Our first example of a paradox is perhaps the oldest and most well-known of all paradoxes. It is called the "liar's paradox." Consider the statement

This statement is false.

If this statement is in fact false, then it is false that this statement is false; thus, this statement is true. Hence, if the statement is false, then it is true. On the other hand, if the statement is true, then it is true that the statement is false; thus the statement is false. We see that if the statement is false, then it is true; and if it is true, then it is false.

The liar's paradox can be explained by taking the position that the sentence upon which the paradox is built—the sentence "This statement is false"—is not really a statement at all. Recall that by agreement a statement is a sentence that is either true or false but not both. The liar's "statement" is neither true nor false, and so it is not a statement. We can take the position that the liar's statement is just nonsense and let it go at that.

Another ancient paradox involves a shifty crocodile and a clever father. The crocodile has stolen a child and has been hunted down by the child's father. The crocodile agrees to return the child to his father if the father can correctly guess whether or not the crocodile really will return the child. What should the crocodile do if the father guesses that the crocodile will not return the child?

The crocodile's agreement is suspect because this agreement seems not really to be an agreement at all. The sentence "I will return your child to you if you can guess whether or not I will return your child to you" may, like the liar's sentence, be regarded as simply nonsense.

Each of the liar's and the crocodile's sentences are constructed according to apparently proper language rules using seemingly appropriate words, but each of these sentences just doesn't make sense. That is, neither of these sentences can really be a statement at all. If either were a statement, then we would have a logical contradiction; and our logic does not permit such logical contradictions. These apparent statements are not really statements in the logical sense because a statement must be either true or false. So the liar's and the crocodile's paradoxes are just apparent paradoxes.

Next let us consider the situation in the Foosland, Illinois, Public Library. This library, like all libraries, contains many catalogs of books. There is the card catalog itself, and there are various bibliographies that are really nothing more than catalogs of books. For example, there is a catalog of the complete works of Alfred Lord Tennyson. Now this catalog obviously was not written by Lord Tennyson himself, so the catalog would not be listed in itself. But the catalog of books published in 1960, publication date December 31, 1960, obviously would list itself. That is, this catalog of books was itself published in 1960, so it would list itself. Certainly, most catalogs would not list themselves, and these are by far the most common kinds of catalog. For identification purposes, let us agree to call such a catalog an *ordinary* catalog. An ordinary catalog, then, is one that does not list itself. The other catalogs, like the catalog of books for 1960, shall be called *special* catalogs. A special catalog is one that lists itself. It should be clear that every catalog is either ordinary or special, but not both.

As the librarian at the Foosland Public Library likes to catalog things, he undertakes to compile a catalog of all the ordinary catalogs in his library. We shall refer to this catalog as the Foosland Catalog. It lists all catalogs in the library that do not list themselves. But the existence of this Foosland Catalog raises the question: Is the Foosland Catalog an ordinary catalog or a special catalog? It must be one or the other. You should try to decide which it is before reading further.

If the Foosland Catalog is an ordinary catalog, then according to what we mean by ordinary, the Foosland Catalog does not list itself. Now remember that the Foosland Catalog lists exactly those catalogs that are ordinary. Because the Foosland Catalog is not listed in the Foosland Catalog, the Foosland Catalog is not ordinary—it is special. What we have just proved is that if the Foosland Catalog is ordinary, then it is special!

On the other hand, if the Foosland Catalog is special, then it does list itself (this is what we mean by a special catalog). So the Foosland Catalog is listed in the Foosland Catalog. But every catalog listed in the Foosland Catalog is an ordinary catalog. Hence the Foosland Catalog must be ordinary too. This proves that if the Foosland Catalog is special, then it is ordinary!

We have shown that if the Foosland Catalog is ordinary, then it is special and if it is special then it is ordinary. Something, clearly, is wrong here. This paradox is sometimes called the "librarian's paradox."

When we strip this paradox of its linguistic trappings, we see that it is not really a paradox in the field of librarianship, it is really a paradox in the theory of sets. The catalogs are really nothing more than sets of objects, the ordinary catalogs are sets that do not contain themselves as one of their own objects, and the special catalogs are sets that do contain themselves as one of their own objects. The Foosland Catalog is simply the collection (synonym for set) of all sets that do not contain themselves as one of their own objects. What the paradox shows, in set-theoretic language, is that there is no such thing as the collection of all sets that do not contain themselves as one of their own objects. It is not that this collection is the empty set or any other such thing—to speak of this collection is simply and purely nonsense. In the language of the paradox itself, it is completely meaningless to talk about the catalog of all ordinary catalogs. But, you say, how can it be meaningless to talk about this catalog—didn't you just do it? No. The phrase

The catalog of all ordinary catalogs

is just as meaningless a jumble of words as is the phrase

Cow hypotenuse swatch spigot horseshoe.

Bertrand Russell (English, 1872–1970) communicated a paradox like this one to the German mathematician Gottlob Frege (1848–1925) in 1902. Frege had just completed a huge work on arithmetic, his purpose being to restructure all of arithmetic upon the "firm logical basis" of Cantor's set theory. The Russell paradox came as a real blow to Frege because it effectively negated his just-finished work. The paradox, in effect, asserted that Cantor's set theory was capable of giving rise to contradictions, and mathematicians abhor contradictions as nature used to abhor a vacuum. The subsequent revision of Cantor's set theory so that such paradoxes would be avoided is only now being completed.

Exercises 4.2

1. Referring to the example of the Foosland Library, can you think of any other catalogs beside the catalog of books published in 1960 that would list themselves? (You can make them up—they do not really have to exist.)

2. The expression, "The least whole number not namable in fewer than 22 syllables" names a certain whole number. Since this whole number cannot be named in fewer than 22 syllables, the expression above, which names this whole number, must contain not fewer than 22 syllables. Comment.

3. There once was a town with a law that every male inhabitant of the town had to be freshly shaved each Sunday morning. But the town had only one barber, who was hard pressed to shave all the men in just one morning. So another law was passed to the effect that the men who shaved themselves regularly had to shave themselves on Sunday morning, too. The barber could shave those and only those men who never shaved themselves. This law worked and presented no problems and all of the people in the town were happy. What was the sex of the barber? Why?

4. Call an adjective *autological* if it describes itself. For example, the adjective "short" is autological because it describes itself. So are "polysyllabic" and "English." Call an adjective *heterological* if it does not describe itself. Examples are "long," "monosyllabic," and "French." Is the adjective "heterological" heterological or autological?

5. Let $x = 1$. Then, since $1^2 = 1$, it is true that $x^2 = x$. Then $x^2 - 1 = x - 1$. Now, since $x^2 - 1 = (x - 1)(x + 1)$, we have $(x - 1)(x + 1) = (x - 1)(1)$, from which it follows that $x + 1 = 1$ by the cancellation property. So $x = 0$. That is, $1 = 0$. This paradox can be explained. Can you do it?

6. The tortoise was given a headstart in his race with Achilles. Poor Achilles—he never could catch that tortoise after that. For each time Achilles reached a spot where the tortoise had been, the tortoise had moved on a certain distance. Thus Achilles was continually reaching places where the tortoise had already been; thus he could never get ahead of the tortoise. Moral: The pursuer can never catch the pursued. [This paradox was given by Zeno (Greek, *c.* 450 B.C.).] Study this paradox.

4.3 Relations on Sets

Now that we have identified and discussed the objects of the mathematical system of sets, we may define some relations. One of the relations on sets is the relation of **equality of sets.** We say that two sets A and B are equal if they contain exactly the same objects. For example, the set of all whole numbers less than 10 is equal to the set of all whole numbers whose squares are less than 82. When two sets A and B are equal, we write $A = B$.

Example 1. The set $\{2, 3\}$ has exactly the same objects as the set $\{3, 2\}$—each set contains the numbers 2 and 3 and no other objects. Hence $\{2, 3\} = \{3, 2\}$. This example illustrates that the order in which the objects of a set are listed is irrelevant.

Example 2. The sets $\{2, 3\}$ and $\{3, 2, 2, 3\}$ are equal, since they each contain the whole numbers 2 and 3 and no other objects. This example illustrates that in listing the objects of a set, it makes no difference whether an object is listed more than once. However, it is convenient to adopt the convention that we shall not repeat the listing of an object.

Another relation on sets is the **proper subset** relation. The sets $\{2, 3\}$ and $\{1, 2, 3\}$ are related by this relation. That is, the set $\{2, 3\}$ is a proper subset of the set $\{1, 2, 3\}$. More generally, to say that set A is a proper subset of set B means that every object of A is also an object of B and B contains at least one object not contained in A. The set $\{a, b, c\}$ is a proper subset of the set $\{z, a, b, d, c\}$, since each object of $\{a, b, c\}$ is also an object of $\{z, a, b, d, c\}$ and $\{z, a, b, d, c\}$ contains at least one object that $\{a, b, c\}$ does not. When A is a proper subset of B, we write $A \subset B$.

The relation of proper subset lends itself to illustration by pictures. In Figure 4.1 we have illustrated two sets A and B such that $A \subset B$.

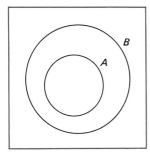

Figure 4.1

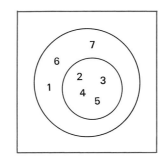

Figure 4.2

Example 3.   The sets $\{2, 3, 4, 5\}$ and $\{1, 2, 3, 4, 5, 6, 7\}$ are illustrated in Figure 4.2.

A third relation on sets is the **subset** relation. When we say that a set A is a subset of a set B, we mean that either A is equal to B or that A is a proper subset of B. The symbol for the subset relation is $\subseteq$.

Example 4.
$$\{1, 2, 3\} \subseteq \{1, 2, 3, 4\} \qquad \{1, 2, 3\} \subseteq \{1, 2, 3, 4, \ldots\}$$
$$\{1, 2, 3\} \subset \{1, 2, 3, 4\} \qquad \{1, 2, 3\} \subset \{1, 2, 3, 4, \ldots\}$$
$$\{1, 2, 3\} \subseteq \{1, 2, 3\} \qquad \{1, 2, 3, \ldots\} \subseteq \{1, 2, 3, 4, \ldots\}$$
$$\{1, 2, 3\} \not\subset \{1, 2, 3\} \qquad \{1\} \not\subseteq \{2\}.$$

The *subset* relation is a combination of the *equality* and *proper subset* relations in the same way that the *is less than or equal to* relation for whole numbers is a combination of the *is equal to* and *is less than* relations.

A fourth relation on sets is the **disjointness** relation. The sets A and B are called disjoint if they contain no objects in common.

Example 5.   The set E of all even whole numbers and the set O of all odd whole numbers are disjoint sets. The set of all one-digit numbers is disjoint from the set of all two-digit numbers.

Figure 4.3 illustrates a pair of disjoint sets A and B.

There is a fifth relation on sets called the "relation of being in one-to-one correspondence with," which we shall consider in Section 4.5.

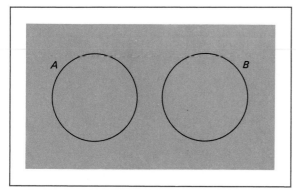

Figure 4.3

Exercises 4.3

1. Which of the following are proper subsets of $\{1, 2, 3\}$? Which are disjoint from $\{1, 2, 3\}$? Which are equal to $\{1, 2, 3\}$? Which are subsets of $\{1, 2, 3\}$?
   (a) $\{1\}$. (b) $\{1, 3\}$. (c) $\varnothing$.
   (d) $\{1, 3, 4\}$. (e) $\{3, 1, 2\}$. (f) $\{2, 6\}$.
   (g) $\{5, 6\}$. (h) $\{3, 1, 5\}$.

2. List all the subsets of the set $\{1\}$. Of $\{1, 2\}$. Of $\{1, 2, 3\}$. Can you make a conjecture as the number of subsets of a set with n objects?

3. A subset of a set S that is not a proper subset is called an *improper* subset. If S is a set, how many improper subsets does S possess?

4. As in Figure 4.1, illustrate sets A, B, and C such that
   (a) $A \subset B$ and $B \subset C$. (b) $A \subset C$, $A \subset B$, but $B \not\subset C$.

5. As in Figure 4.2, illustrate the following sets.
   (a) $\{1, 2, 3, 4, 5, 6, 7\}$, $\{3, 5, 6\}$, and $\{1, 2, 3\}$.
   (b) $\{1, 2, 3\}$, $\{1, 2\}$, $\{1\}$, $\{2\}$, $\{2, 3\}$, $\{3\}$, and $\{1, 3\}$.

6. Find a subset of all digits (the whole numbers 0 through 9) that is disjoint from the set $\{1, 2, 3, 4, 5, 6, 7\}$. How many subsets are there?

7. Is it possible to find a set that is disjoint from one of its subsets? (Careful!)

8. Complete the following table with *Yes* and *No*.

	Reflexive	Symmetric	Transitive
Equality of sets			
Subset			
Proper subset			
Disjointness			

*9. If we have a collection of two or more sets $S, T, U, V, \ldots$, then we say that these sets are **pairwise disjoint** if every two of these sets are disjoint. Are the sets $\{1\}$, $\{1, 2\}$, $\{3\}$, and $\{4, 5, 6\}$ pairwise disjoint? Find a collection of six sets of whole numbers that are pairwise disjoint.

*10. Let us use the (nonstandard) term *overlaps* to describe the relationship between a set A and a set B when neither set is a subset of the other but the sets are not disjoint. For example, $\{1, 2, 3\}$ overlaps $\{2, 3, 4\}$. Draw a picture to illustrate the general relationship between sets A and B such that A overlaps B. Is this relation reflexive? Symmetric? Transitive?

11. Construct a flow chart for determining whether a finite set A is a subset of a set B.

12. Which of the following are logically equivalent?
    (a) A is either equal to B or A is a proper subset of B.
    (b) A is disjoint from B.
    (c) A is a subset of B.
    (d) A and B have no objects in common.

(e) A is not equal to B.

(f) A is not a subset of B or B is not a subset of A.

(g) A contains no object of B and B contains no object of A.

4.4 Operations on Sets

Having defined some relations on sets, we are now in a position to define some binary operations on sets. The operations that we shall discuss here are familiar to the reader already, although he may not have used these ideas in quite the same way that we shall use them.

The first operation on sets is related to the familiar operation of addition of whole numbers. The operation is called **union** and is performed on two sets A and B by combining the objects of the two given sets to form a new third set. For example, the result of the union operation applied to the sets $\{1, 2\}$ and $\{3, 4, 5\}$ is the set $\{1, 2, 3, 4, 5\}$. The symbol that denotes this operation is the symbol $\cup$. So, we would write $\{1, 2\} \cup \{3, 4, 5\} = \{1, 2, 3, 4, 5\}$. Here are more examples.

Example 1.   The union of the sets $\{1, 2, 3\}$ and $\{2, 3, 4\}$ is the set $\{1, 2, 3, 4\}$. We could have written $\{1, 2, 3\} \cup \{2, 3, 4\} = \{1, 2, 3, 2, 3, 4\}$, but we have agreed not to repeat any object in the listing of the objects of a set.

Example 2. $\{1, 2, 3\} \cup \{0, 1, 2, 3, 4, 5, 6\} = \{0, 1, 2, 3, 4, 5, 6\}$. This example illustrates that if A is a proper subset of B, then the union of A and B is B.

Example 3. $\{a, b, c\} \cup \varnothing = ?$

Solution: Since the empty set contains no objects at all, this union is simply the set $\{a, b, c\}$.

Figure 4.4 shows the union of two arbitrary sets A and B. The shaded region represents $A \cup B$.

The second binary operation on sets that we shall need is called the **intersection** operation. The intersection of two streets can be thought of as that portion of the earth which lies in both streets at the same time—that portion of the earth common to both streets. If A and B are sets, then we say that their intersection is the set consisting of all objects common to both sets. For example, the intersection of $\{2, 3, 4\}$ and $\{3, 4, 5\}$ is the set $\{3, 4\}$. We use the special symbol $\cap$ to denote this operation, and so we can write $\{2, 3, 4\} \cap \{3, 4, 5\} = \{3, 4\}$. Here are more examples.

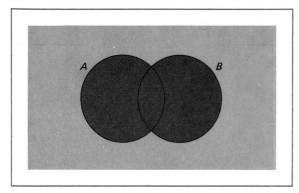

Figure 4.4

Example 4. $\{a, b, c\} \cap \{a, b, c, d, e\} = \{a, b, c\}$. This example illustrates that if A is a proper subset of B, then the intersection of A and B is the set A.

Example 5.   If A and B are disjoint sets, then what is $A \cap B$?

Solution: Since disjoint sets have no objects in common, $A \cap B = \emptyset$. The intersection of two arbitrary sets A and B is pictured in Figure 4.5 as the shaded portion of that illustration.

Exercises 4.4 _____

1. By using the sets $A = \{1, 2, 3\}$, $B = \{2, 3, 4\}$, and $C = \{0, 5, 6\}$, describe the following sets by listing their objects.
   (a) $A \cap (B \cup C)$.   (b) $A \cap (B \cap C)$.
   (c) $A \cup (B \cup C)$.   (d) $(A \cap B) \cup (A \cap C)$.

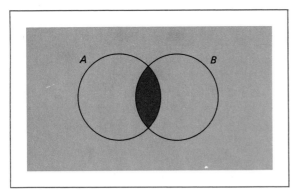

Figure 4.5

2. If $A \subset B$ and $B \subset C$, then $A \cap B = ?$ $A \cap C = ?$ $B \cap C = ?$
3. If $A \subset B$ and $B \subset C$, then $A \cup B = ?$ $A \cup C = ?$ $B \cup C = ?$
4. If $A \cup B = B$ and $A \cap B = \emptyset$, what can you say about A?
5. Which of these statements are logically equivalent?
     (a) $A \subseteq B$.   (b) $B \subseteq A$.   (c) $A \cup B = A$.   (d) $A \cap B = A$.
6. Are the operations of union and intersection commutative? Are they associative? Give some examples.
7. Can you find three sets A, B, and C such that $A \cup C = B \cup C$ and $A \neq B$? What does your answer tell you about the union operation? Can you find three sets such that $A \cap C = B \cap C$ and $A \neq B$? What does this tell you about the intersection operation?
8. By using the sets $A = \{1, 2, 3, 4\}$, $B = \{2, 3, 6, 7\}$, and $C = \{2, 4, 6, 8\}$, verify that the operations of union and intersection share the following two distributive properties.
     (i) $A \cap (B \cup C) = (A \cap B) \cup (A \cap C)$.
     (ii) $A \cup (B \cap C) = (A \cup B) \cap (A \cup C)$.
   These two properties show that each of union and intersection distributes over the other (on the left).
9. The kingdoms of Klutz and Smorg both claim sovereignty over the Duchy of Muck. The population of Muck is 13. Klutz has a population of 7891 and Smorg has a population of 9047. What is the combined population of Klutz and Smorg?
*10. Is there an identity object for the operation of union? That is, is there a set I such that $A \cup I = I \cup A = A$ for all sets A? Does the operation of intersection have an identity? (You should assume that we are asking these questions relative to some universal set U. See Exercise 9, Section 4.1, for a discussion of universal sets.)
*11. If A and B are sets, then the set of all objects in A that are not in B is denoted by $A - B$ and is called the **set difference** of A and B(in that order).
     (a) Find $A - B$ if $A = \{1, 2, 3\}$ and $B = \{2, 3\}$. If $A = \{1, 2, 3\}$ and $B = \emptyset$. If $A = \{1, 2, 3\}$ and $B = \{5, 6, 7\}$.
     (b) If $A - B = A$, how are A and B related?
     (c) If $A - B = \emptyset$, how are A and B related?
     (d) Is the operation of set difference commutative?
     (e) By using circles, illustrate the set $A - B$.
*12. Use the array of circles shown on p. 112 and shade in the region that represents each of the following sets. These diagrams are called *Venn diagrams*.
     (a) $(A \cap B) \cap C$.
     (b) $(A \cup B) \cap C$.
     (c) $B \cup C$.
     (d) $(A \cap C) \cup (B \cap C)$.
     (e) $A \cap (B \cap C)$.
     (f) $B \cup (C \cap A) \cup C$.
   By comparing the shaded regions, can you find any equalities between sets in this list?

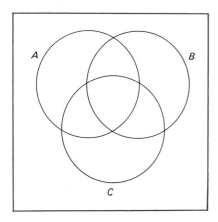

*13. In a certain high school the total enrollment in chemistry is 60, in biology 75, and in physics 70. (A single student enrolled in two courses would be counted twice.) Six students are enrolled in all three courses; 20 students are enrolled in chemistry and biology; 38 students are enrolled in chemistry and physics; and 29 students are enrolled in biology and physics. By using the Venn diagram in Exercise 12 to help you, determine exactly how many students we are talking about.

*14. Zero is called an annihilator for multiplication in the system of whole numbers because $x \cdot 0 = 0 \cdot x = 0$ for all whole numbers x. Is there an annihilator for the operation of intersection in the theory of sets? That is, is there a set A such that $X \cap A = A \cap X = A$ for all sets X? Is there an annihilator for union?

15. Construct a flow chart showing how you would test to determine whether or not an element x belonged to the union of two sets A and B.

16. Construct a flow chart showing how you would test to determine whether or not an element x belonged to the intersection of two sets A and B.

4.5   One-to-One Correspondence

We have discussed the objects, some of the relations, and some of the operations of the mathematical system of set theory. In this section we shall discuss one more relation, a relation that is prerequisite to the study of "infinite numbers."

The objects of the sets $\{a, b, c\}$ and $\{1, 2, 3\}$ can be paired as shown in Figure 4.6. This pairing is one that pairs each object of each set with exactly one object of the other set. Such a pairing between the objects of the two sets is called a **one-to-one correspondence.** More generally, two sets A and B are said to be in one-to-one correspondence if the objects of the two sets are paired in such a way

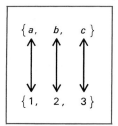
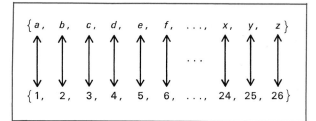

Figure 4.6   **Figure 4.7**

that each object of A is paired with exactly one object of B and each object of B is paired with exactly one object of A.

Here are some more examples of pairs of sets that have been placed in one-to-one correspondence.

Example 1. The set of letters of the alphabet and the set of nonzero whole numbers less than 27 can be placed in a one-to-one correspondence, as shown in Figure 4.7. The two sets are related to each other by the relation *is in one-to-one correspondence with*.

Example 2. There are six different ways to construct a one-to-one correspondence between the sets $\{1, 2, 3\}$ and $\{a, b, c\}$. These six different pairings are shown in Figure 4.8.

Examples 1 and 2 dealt with finite sets. More interesting things happen when we put infinite sets in one-to-one correspondence.

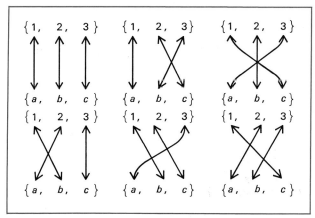

Figure 4.8

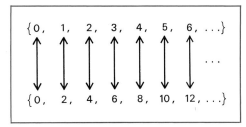

Figure 4.9

Example 3. The set of all whole numbers can be put into one-to-one correspondence with the set of all even whole numbers, as shown in Figure 4.9.

Example 3 demonstrates that an infinite set may be put into one-to-one correspondence with a proper subset of itself. This cannot happen with finite sets. That is, a finite set cannot be put into one-to-one correspondence with a proper subset of itself. (Try to construct a one-to-one correspondence between the set $\{1, 2, 3\}$ and its proper subset $\{1, 2\}$.) This idea was taken by Cantor to be the defining characteristic of an infinite set. That is, we define an **infinite set** to be one that can be put in one-to-one correspondence with a proper subset of itself, and we define a **finite set** to be one that cannot be put into one-to-one correspondence with any of its proper subsets.

Prior to 1872 it had been thought that the fact that some sets could be put into one-to-one correspondence with proper subsets of themselves was paradoxical. Indeed, if you insist that all sets have the same properties and obey the same laws as do finite sets, then the fact that some sets (the infinite ones) can be put into one-to-one correspondence with proper subsets of themselves is indeed paradoxical. But Cantor (and others as well) saw that this was not paradoxical at all—it was simply the characteristic of infinite sets that differentiates them from finite sets and hence could be taken as the definition of an infinite set. This was the first breakthrough toward a theory of infinite sets.

Example 4. Explain why the set of whole numbers greater than 5 is an infinite set.

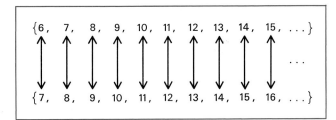

Figure 4.10

Solution: By definition of infinite set, the set of whole numbers greater than 5 is infinite because it can be put into one-to-one correspondence with one of its proper subsets. It can, for instance, be put into one-to-one correspondence with its proper subset {7, 8, 9, 10, 11, . . .}, as shown in Figure 4.10.

Example 5.   Consider the parallel line segments $\overline{AB}$ and $\overline{CD}$ shown in Figure 4.11. Each of these line segments is a set of points, and as sets these two segments can be placed in one-to-one correspondence in the following manner. Construct the lines $\overleftrightarrow{CA}$ and $\overleftrightarrow{DB}$ in order to determine the point P. Then pair points X on line segment $\overline{AB}$ and Y on line segment $\overline{CD}$ by drawing lines through P as shown. In this way each point of $\overline{AB}$ can be paired with exactly one point of $\overline{CD}$ and, conversely, each point of $\overline{CD}$ can be paired with exactly one point of $\overline{AB}$.

Example 6.   Let T denote the set of all true statements and F denote the set of all false statements. How do the number of statements in T compare with the number of statements in F? Are there more, fewer, or the same number?

Solution: There are exactly the same number of true statements as there are false statements. We can prove this by establishing a one-to-one correspondence between the elements of the sets T and F. Let $s \in T$. Then $\sim s$ is false (since s is true) and so $\sim s \in F$. Conversely, if $s \in F$, then $\sim s \in T$. The correspondence

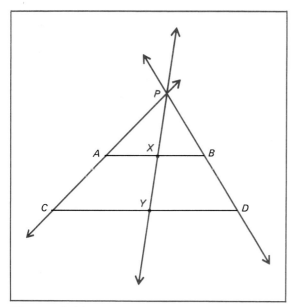

Figure 4.11

just described is a one-to-one correspondence between the true statements and the false statements.

In the next section we are going to use the relation of *is in one-to-one correspondence with* to define finite and infinite cardinal numbers.

Exercises 4.5 —————————————————————————————

1. Illustrate all possible one-to-one correspondences between the following sets.
   (a) $\{a, b\}$ and $\{1, 2\}$.   (b) $\{a, b, c, d\}$ and $\{1, 2, 3, 4\}$.
   (c) $\{a, b, c\}$ and $\{1, 2\}$.
2. Set up a one-to-one correspondence between the set of all whole numbers and the set $\{0, 3, 6, 9, 12, 15, 18, 21, \ldots\}$. Under your correspondence, which multiple of 3 would correspond to the whole number 145? To which whole number would the multiple 333 correspond?
3. Explain why the set of all whole numbers that are multiples of 3 is an infinite set.
4. Discuss the bigamy laws in terms of one-to-one correspondence.
5. What does the fact that your college has chosen to use a "number-name" of one kind or another to designate you, rather than your given name, have to do with the idea of one-to-one correspondence?
*6. We call two sets A and B that can be placed in one-to-one correspondence equivalent sets. For example, $\{a, b, c\}$ and $\{1, 2, 3\}$ are equivalent sets.
   (a) Are equivalent sets necessarily equal sets? Are equal sets necessarily equivalent sets?
   (b) Which of the properties of reflexivity, symmetry, and transitivity are possessed by the relation *is equivalent to*?
7. Let $\mathcal{C}_1$ and $\mathcal{C}_2$ be circles of radius 1 inch and 2 inches, respectively. These circles may be regarded as being sets of points in the plane. Show that these sets of points can be put into one-to-one correspondence with each other. (*Hint:* Situate the circles so that their centers coincide. Then draw a half-line issuing from this center. The intersection points can be paired. Explain why this method gives a one-to-one correspondence.)

4.6†   Cardinal Numbers

In this section we are going to introduce some of Cantor's "infinite numbers." These numbers are obtained from infinite sets in the same way that the whole numbers are obtained from finite sets. We shall begin, therefore, by considering finite sets.

Every finite set contains a certain definite number of objects. The set $\{a, b, c\}$

contains 3 objects. We call the number of objects in a finite set the **cardinal number** of that finite set. Clearly, the cardinal number of a finite set is a whole number. When a whole number is being used to indicate the number of objects in a finite set, we say that the whole number is being used as a cardinal number. When we say that there are 26 letters in the English alphabet, we are using the whole number 26 as a cardinal number.

There is another way to use whole numbers. For example, when we say that z is the 26th letter of the alphabet, we are using 26 to indicate the position of the letter z in an ordering of the letters of the alphabet. In this case we say that we are using the whole number 26 as an **ordinal number.**

Example 1.   The page number of this page refers to the position of this page in an ordering of pages. Hence the number of this page is an ordinal number. It is a whole number being used in the ordinal sense.

Example 2.   The set $\{a, b, c\}$ has cardinal number 3. $\{a, b, c\} \cup \{a, b, d\}$ has cardinal number 4. $\{1, 2, 3, \ldots\}$ does not have a *finite* cardinal number.

It is at this point that the concept of one-to-one correspondence becomes important. For what does is mean to say that two finite sets A and B have the *same cardinal number?* It means that the finite sets A and B can be placed into one-to-one correspondence with each other. That is, *the cardinal number of set A is equal to the cardinal number of set B if and only if A can be placed into one-to-one correspondence with B.*

We often talk of "counting" the number of objects of a finite set. What exactly do we do when we "count" a finite set? Suppose we look at the set $\{a, b, c\}$ and "count" its objects. We "count" this set by looking at a and thinking 1:

$$a \leftrightarrow 1.$$

Then we look at b and think 2:

$$b \leftrightarrow 2.$$

Finally, we look at c and think 3:

$$c \leftrightarrow 3.$$

What we have done is to construct a one-to-one correspondence between the given set and the set $\{1, 2, 3\}$, and so we say that the given set contains 3 objects—has cardinal number 3. The counting process is, therefore, seen to be nothing more than a process that sets up a one-to-one correspondence between a given finite set and a set of the special form $\{1, 2, 3, 4, 5, \ldots, n\}$. After this correspondence has been constructed, we conclude that the given set has cardinal number n.

Example 3. On the first day of class a clerk enters a classroom and asks the teacher how many students she has in the room. The teacher scans the room and observes that every seat is occupied and that there are no students standing or sitting on the floor. She knows that there are 30 seats in the room. Without actually counting the number of students, she replies that there are 30 students in the room. The teacher has used the fact that the set of students and the set of chairs are in one-to-one correspondence; and, since one of these sets contains 30 elements (has cardinal number 30), then the other must also.

Example 4. It is interesting that most children will at some stage of their intellectual development believe that jar *A* in Figure 4.12 has more marbles than jar *B*. These children have not yet grasped the principle that two sets that are in one-to-one correspondence have the same number of elements. They are basing their comprehension of "number size"—that is, cardinal number—upon recognized quantitative measures that they already comprehend, such as volume or dimension. In this instance the volume of the jar is being used.

We have seen that every finite set has a cardinal number and that two finite sets have the same cardinal number if and only if they are in one-to-one correspondence with each other. Does this mean that only finite sets have cardinal numbers? Not at all. Cantor's primary contribution was to understand that every set, finite *or* infinite, could have a cardinal number and that the cardinal numbers of two sets, finite *or* infinite, were equal if and only if the two sets were in one-to-one correspondence with each other. Of course, the cardinal number of an infinite set could not be a whole number; it would have to be some new kind of number—a new kind of number that we call an **infinite cardinal number.** For example, the set *W* of all whole numbers is an infinite set and thus has a cardinal number which is an infinite cardinal number. We denote the cardinal number of this set

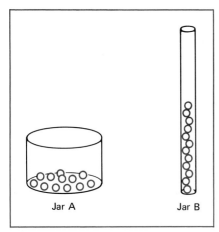

Jar A   Jar B

Figure 4.12

by the symbol $\aleph_0$ (read "aleph-sub-nought"). The cardinal number $\aleph_0$ is an infinite cardinal number.

We have one example of a set of cardinality $\aleph_0$. Can you think of any others? Remember that two sets have the same cardinal number if and only if they can be placed into one-to-one correspondence with each other. Therefore, any set that can be put into one-to-one correspondence with the set W of whole numbers will also have cardinal number $\aleph_0$. Well, we have seen that the set E of even whole numbers can be put into one-to-one correspondence with W; thus, E also has the cardinal number $\aleph_0$.

This situation is one that bothers many students at first, so perhaps we should discuss it more fully. We have here two sets W and E, one of which is a proper subset of the other; yet they have the same cardinal number. This sort of thing cannot happen with finite sets. That is, if A and B are finite sets and if $A \subset B$, then the cardinal number of A must be less than the cardinal number of B. But if A and B are infinite sets, then simply because B has objects not contained in A does not necessarily mean that A and B have different cardinal numbers. The reason for this is that infinite sets can be placed into one-to-one correspondence with proper subsets of themselves.

Cantor was not the first to see clearly what an infinite set was, but he was the first to see that there were infinite sets having different cardinal numbers. In other words, $\aleph_0$ is not the only infinite cardinal number; it is only the smallest of infinitely many different infinite cardinal numbers. A larger infinite cardinal number is $\aleph$, which is the cardinal number of the set of all points on a line. Hence $\aleph > \aleph_0$.

Example 5. Suppose we construct two line segments of different lengths, as shown in Figure 4.13. Each of these segments is in actuality a set of points; that is, these segments are sets. As sets they have cardinal numbers. These cardinal numbers must be infinite for there are surely infinitely many points on each of the line segments. How do the cardinal numbers of these two sets compare?

Solution: They are equal. The reason for this is that these two line segments can be placed in one-to-one correspondence with each other and so *must* have the same cardinal numbers. In fact, we did this in Example 5 of Section 4.5.

Example 6. Consider the two circles in Figure 4.14. The radius of circle $\mathcal{C}_1$ is only half the radius of circle $\mathcal{C}_2$. Like the line segments in the example above,

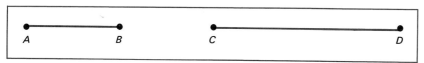

A   *B*   *C*   *D*

Figure 4.13

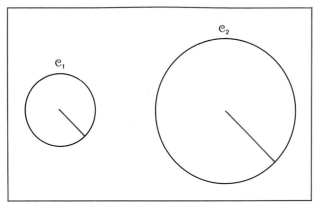

Figure 4.14

these circles are sets of points in the plane. As sets, each circle has a cardinal number. How do their cardinal numbers compare?

Solution: Perhaps you have already guessed that they are equal. This is true, and we can most easily prove this by setting up a one-to-one correspondence as follows. Place circle $\mathcal{C}_1$ and circle $\mathcal{C}_2$ so that their centers coincide. Refer to Figure 4.15. By drawing half-lines issuing from the centers of the circles, we can pair up the points of the two circles in a one-to-one correspondence. For example, as Figure 4.15 shows, the points N and M correspond as do the points K and L. Given a point P on circle $\mathcal{C}_1$, you can find the point on circle $\mathcal{C}_2$ to which it corresponds by drawing a half-line issuing from the center of $\mathcal{C}_1$ and passing through the point P. The point on circle $\mathcal{C}_2$ where this half-line cuts circle $\mathcal{C}_2$ is the point on $\mathcal{C}_2$, say Q, that corresponds to P.

It must be clear that the correspondence so determined is one-to-one. That is, to each point on circle $\mathcal{C}_1$ there corresponds exactly one point on circle $\mathcal{C}_2$, *and*

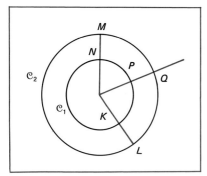

Figure 4.15

to each point on circle $\mathcal{C}_2$ there corresponds exactly one point on circle $\mathcal{C}_1$. These two circles contain exactly the same number of points even though one of them has a greater circumference than the other.

Infinite cardinal numbers even larger than $\aleph$ can be constructed by using the following scheme: If S is a set, we call the collection of all subsets of S the **power set** of S and denote it by $P(S)$. Cantor proved that the cardinal number of the set $P(S)$ is always greater than the cardinal number of S. Thus W, $P(W)$, $P(P(W))$, $P(P(P(W)))$, $P(P(P(P(W))))$, $P(P(P(P(P(W)))))$, . . . , are infinite sets, each of which has cardinal number greater than the one that came before it. Hence there are infinitely many infinite cardinal numbers. We shall not prove this theorem, but let us consider an example of it when S is a finite set.

Example 7. What is the cardinal number of the power set of a set that has cardinal number 3?

Solution: Consider the set $\{a, b, c\}$. Then the power set contains as its objects the following sets: $\varnothing$, $\{a\}$, $\{b\}$, $\{c\}$, $\{a, b\}$, $\{a, c\}$, $\{b, c\}$, and $\{a, b, c\}$. Hence the power set has cardinal number 8.

Example 7 illustrates the theorem that if S is a finite set of cardinal number n, then the power set of S is a finite set whose cardinal number is 2^n.

If you are left up in the air by this discussion of infinite cardinal numbers, you might be able to better understand the situation in which a member of one of the primitive tribes that count "One—two—many" found himself when first confronted by a missionary who talked to him about the 12 disciples and the 10 commandments. His experience had not prepared him to accept such numbers— much less to comprehend the "negative" numbers used to name dates before the birth of Christ—as intuitively reasonable ideas. In something of the same way, your intuition has not prepared you to discuss infinite sets and their cardinal numbers because you have never dealt on an intuitive level with any but finite sets. For this reason, the study of such numbers (and we shall work with them only a little in this book) must necessarily be conducted on an abstract level. Only after extensive association with infinite sets can one claim any intuition concerning them. In fact, there is a kind of mathematician-logician-philosopher who claims that it is impossible to regard the notion of an infinite set as a legitimate concept in mathematics—or in·anything else for that matter. Such logicians reject the use of the infinite altogether. The reason that they are in a minority is that, generally speaking, mathematicians want to get results with their mathematics; they feel that certain results must be obtainable and these results are not obtainable if one restricts himself by denying the use of infinite sets. Doing modern mathematics without infinite sets is like trying to swim in a lead life jacket.

1. What is the cardinal number of the set $\{1, 10, 100, 1000, 10000, \ldots\}$? Show that your answer is correct by using the notion of one-to-one correspondence.
2. How many different ways are there to count the number of objects in the set $\{a, b, c\}$? In the set $\{a, b, c, d\}$? In a set of n objects?
3. In each sentence identify whether the whole number is being used as a cardinal number or as an ordinal number.
   (a) Ugh; I gained another pound.
   (b) I'll meet you at 6 sharp.
   (c) You owe me $6.
   (d) There sure are a lot of rooms in this building—there is room 24516798013.
   (e) I shot three on the third hole, four on the fourth hole, five on the fifth hole, six on the sixth hole, threw my eight clubs in the water hazard on the seventh hole, and was first into the clubhouse after 10 A.M.
4. Find three sets of numbers A, B, and C such that $A \subset B \subset C$ and all of these sets have the same cardinal number. Can you find such an example when any one of the sets is finite?
5. Let A and B be two sets, each of which has cardinal number $\aleph_0$. Write $A = \{a_0, a_1, a_2, a_3, \ldots\}$ and $B = \{b_0, b_1, b_2, b_3, \ldots\}$, where the subscripts are used to indicate to which whole number the object corresponds. Set up a one-to-one correspondence between $A \cup B$ and the set of whole numbers in order to show that $A \cup B$ also has cardinal number $\aleph_0$.
6. Consider a square of side 1 inch and the circle of radius $\frac{1}{2}$ inch inscribed inside it. Show that these two geometric figures, when regarded as sets of points, have the same cardinal number. (*Hint:* You might reconsider Example 6.)
7. Let S be a set containing two objects. Determine the cardinal numbers of the power sets $P(S)$, $P(P(S))$, and $P(P(P(S)))$.
8. The usual checkerboard consists of eight rows and eight columns comprising a total of $8 \cdot 8 = 64$ squares. Imagine a checkerboard consisting of $\aleph_0$ rows and $\aleph_0$ columns like that suggested in the drawing. By using the concept of one-to-one correspondence, explain why there are $\aleph_0$ squares in the first row of this checkerboard. How many squares in the first column? How many squares are there altogether? (To answer this question using one-to-one correspondences, use the hint suggested by the whole numbers that have been put into some of the squares.) How many squares in color are there? How many white squares?

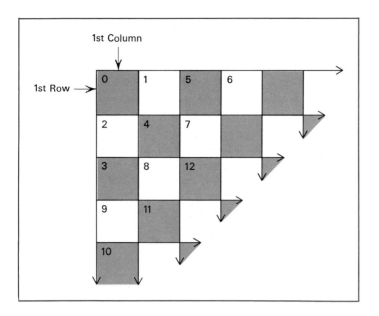

1st Column

1st Row →

0	1	5	6
2	4	7
3	8	12
9	11
10

$4.7^{\dagger}$ Addition of Countable Cardinal Numbers

The cardinal numbers $\aleph_0, 0, 1, 2, 3, 4, 5, \ldots$ are called **countable** cardinal numbers. All other cardinal numbers (for example, $\aleph$ and all larger infinite cardinal numbers) are called **uncountable.** In this section we want to extend the idea of addition of whole numbers (that is, finite cardinal numbers) to include the infinite cardinal number $\aleph_0$. To do this, we must first reexamine the addition of whole numbers in a set-theoretic context.

Suppose that we want to add the whole numbers 3 and 4. How would we accomplish this using the ideas of set theory? In fact, how do we teach very young children to add these numbers? We might ask them to think of two sets, one set containing 3 triangles and the other containing 4 squares. We then ask them to form the union of these two sets and then to count the number of objects in the union. The cardinal number of the union will be the sum of the given cardinal numbers. Thus, to add 3 and 4, we must find two disjoint sets A and B such that the cardinal number of A is 3 and the cardinal number of B is 4. The sum of 3 and 4 is then the cardinal number of $A \cup B$. (Note that A and B must be disjoint or else the union will contain fewer than $3 + 4$ elements. For example, if $A = \{a, b, c\}$ and $B = \{a, b, c, d\}$, then $A \cup B$ contains only 4 elements.)

If we use this same idea, how might we add the countable cardinal numbers 1 and $\aleph_0$? We would first find two disjoint sets A and B such that A has cardinal number 1 and B has cardinal number $\aleph_0$. The sets $A = \{0\}$ and $B = \{1, 2, 3, 4, 5, \ldots\}$ will do. Then we form $A \cup B$ and get the set $\{0, 1, 2, 3, 4, 5, \ldots\}$ of all whole numbers. This union obviously has cardinal number $\aleph_0$. Hence the sum of $\aleph_0$ and 1 is $\aleph_0$:

$$\aleph_0 + 1 = \aleph_0.$$

There is a story about a hotel with $\aleph_0$ rooms that illustrates the above addition fact. The rooms are numbered using the positive whole numbers (room 1, room 2, room 3, and so on). During the tourist season this hotel is almost always full, and in particular it is full on a certain day when Mr. Beaumont walks into the hotel and demands a room. The management would like very much to provide him with a room. Can they do it? Yes. How? The management calls each of their $\aleph_0$ guests and requests that each guest should move into the room whose room number is one greater than the number of the room in which they presently are staying. Hence the man in room 456 should move into room 457, and so on. After all this moving has been accomplished, what is the state of the hotel? Every one of the old guests still has a room, but also room 1 is now vacant. Mr. Beaumont can then be put into this room. We have added one more guest to the original $\aleph_0$ guests, and we still have just $\aleph_0$ guests. That is, $\aleph_0 + 1 = \aleph_0$.

With reference to this same hotel, there was once a baseball game between two teams, each of which had $\aleph_0$ players. The first team arrived at the hotel and each player was given a room. Thus the hotel was filled up. Then the second team arrived and asked for rooms. The management was able to find rooms for these additional $\aleph_0$ players. How did they do this? The management simply had each player on the first team move to the room that had a number twice the number of his original room. Thus the player in room 415 would move to room 830, and so on. After the move had taken place, each player of the first team to arrive still had a room and all of the rooms with odd numbers were now vacant. The players on the second team to arrive were then put into these odd-numbered rooms. This story demonstrates the addition fact

$$\aleph_0 + \aleph_0 = \aleph_0.$$

To subtract one *finite* cardinal number from another, we do just about the same thing that we did to add them. For example, to subtract 3 from 5 (as a child might do), we can select a set of five elements and then select a subset of this set that contains three elements. We then consider the third set of elements in the first set but not in the second. Refer to Figure 4.16. By determining the cardinal number of this "subtraction set," we can determine the number $5 - 3$. To subtract a finite cardinal number from $\aleph_0$, we do exactly the same thing.

Example 1. Determine $\aleph_0 - 2$.

Solution: First, we select a set of cardinal number $\aleph_0$; the set of positive whole numbers will do nicely. Then we select a subset of W containing two elements—of cardinal number 2. The subset $\{1, 2\}$ will serve here. Then we "remove" the subset

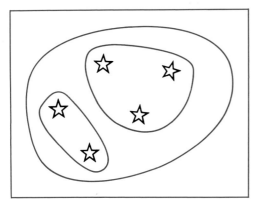

Figure 4.16

from the original set and are left with $\{3, 4, 5, 6, 7, \ldots\}$. We then determine the cardinal number of this remaining set. Now, we can't do this by counting, so we must determine the cardinal number of this set by putting it into one-to-one correspondence with some other set whose cardinal number we already know. You should be able to find a one-to-one correspondence between this "difference set" and the set W. Hence this difference set has $\aleph_0$ elements and so we have proved that $\aleph_0 - 2 = \aleph_0$.

Example 1 demonstrates that if n is any whole number, then $\aleph_0 - n = \aleph_0$. What is $\aleph_0 - \aleph_0$? Here are two solutions to this problem. You should study them carefully and see what you think about each independently.

Solution 1: Consider the set W. This set has cardinal number $\aleph_0$. Now we must select a subset of W having cardinal number $\aleph_0$, and then we shall remove the subset from the whole set and see what remains. We might select E, the set of even whole numbers, as a subset of W having cardinal number $\aleph_0$. Then we would have

$$\{0, 1, 2, 3, 4, 5, \ldots\} - \{0, 2, 4, 6, 8, \ldots\} = \{1, 3, 5, 7, \ldots\}.$$

But the cardinal number of $\{1, 3, 5, 7, 9, \ldots\}$ is $\aleph_0$. Hence

$$\aleph_0 - \aleph_0 = \aleph_0.$$

Solution 2: Again consider W as a set of cardinal number $\aleph_0$. This time select the subset $\{1, 2, 3, 4, 5, 6, \ldots\}$ as the subset of cardinal number $\aleph_0$. Then write

$$\{0, 1, 2, 3, 4, 5, \ldots\} - \{1, 2, 3, 4, 5, 6, \ldots\} = \{0\}.$$

---

and see that since the cardinal number of $\{0\}$ is 1,

$$\aleph_0 - \aleph_0 = 1.$$

Each of these solutions is, taken by itself, acceptable—or apparently so. But taken together they give $\aleph_0 = 1$. Is there some way that we can explain this paradox? Can we eliminate the paradox by restricting what is permissible in set theory?

What we must realize because of the foregoing arguments is that subtraction of infinite cardinal numbers is not an admissible operation on cardinal numbers. We can add infinite cardinal numbers, but we cannot subtract them. Subtraction is an operation on numbers that must invariably produce a single value as the result of any particular subtraction. Because, as we have seen by means of the two solutions to the problem $\aleph_0 - \aleph_0 = ?$, subtraction of infinite cardinal numbers does not invariably produce a single result; and so subtraction is not permitted in the case that both the subtrahend and minuend are infinite cardinal numbers.

This introduction to cardinal arithmetic is the end of the trail for us. We shall want to refer to the cardinal number of certain infinite sets in later chapters, but we shall not become involved with cardinal arithmetic. The point that we have tried to make here is that such a thing as cardinal arithmetic (with infinite cardinal numbers) exists, is meaningful, and is important in mathematics.

Exercises 4.7

1. A hotel has exactly 1000 rooms. On a night when all these rooms are occupied Mr. Beaumont arrives at the hotel and requests a room. What does the manager tell him? How does this situation differ from the one described in the text?
2. Show how to use sets to find the sum of 2 and 5; to find the sum of 5 and $\aleph_0$; and to find the sum of $\aleph_0$ and $\aleph_0$.
3. Prove that $\aleph_0 - 3 = \aleph_0$.
4. The infinitely many roomed hotel is again full when the electrical supply to Wing A, containing 10 rooms, fails. Must the lodgers in these 10 rooms be moved to another hotel, or can a clever desk clerk rearrange all the lodgers in order to accommodate the unlucky 10?
5. By using the interpretation of multiplication by a whole number as repeated addition, it is possible to give meaning to the symbol $n \times \aleph_0$ (where n is a whole number). What meaning can be given to this product?
*6. Use the checkerboard exercise (Exercise 8, Section 4.6) to explain why $\aleph_0 + \aleph_0 = \aleph_0$. (*Hint:* Consider the set of squares in color and the set of white squares.) Explain how the standard checkerboard can be used to explain why $8 \times 8 = 64$. Draw an analogy from this, and, using the checkerboard with $\aleph_0$ rows and $\aleph_0$ columns, could you say what $\aleph_0 \times \aleph_0$ should be equal to?

5

Two Geometries

The term *geometry* is so general as to be almost meaningless. The fact is that there are many different kinds of geometry, and in this chapter we shall look at two of these. We shall begin by considering the old and familiar geometry of Euclid and shall direct our attention toward the relations of congruence and similarity. These relations will be defined in terms of motions through space. Then we shall use these relations to introduce the notion of a geometric invariant. The notion of a geometric invariant will be used to define Euclidean geometry and to demonstrate how other geometries can be defined. Finally, we shall introduce (still using the notion of a geometric invariant) the most important of the geometries different from Euclidean geometry, the geometry called topology.

Throughout this chapter our approach will be entirely intuitive and informal. We shall base our arguments upon pictures and upon facts that we believe you will accept as being intuitively obvious or nearly so. This is in contrast to our study of the system of whole numbers in Chapter 3, where we based our arguments upon a predetermined list of properties possessed by that system.

5.1 Dictionary of Geometric Figures

Before we begin our study of geometry, let us introduce the geometric figures that we shall be working with. The definitions given in this section are for the most part informal and, hence, imprecise. This, however, should cause no difficulty in our application of these ideas. To supply carefully drawn definitions for these terms would, in some cases, be extremely difficult.

In our informal treatment we shall take the notions of **point, line, plane, and space** as undefined terms. We shall assume, then, that we know what these terms mean without defining them. Using these basic ideas (and others as well), we can go on and define other less basic concepts.

By a **geometric figure** we mean nothing more or less than a set of points in space. You are probably more inclined (by reason of experience) to think of geometric figures as having some definite and recognizable shape or configuration, but they may consist of any scattering of points either in the plane (planar figures) or in space (spatial figures). This emphasis on the set nature of geometric figures obviously stems from the relatively recent introduction of set theory into geometry. Euclid, for example, tended to think of geometric figures as entities in their own right rather than as particular sets of points.

Let ℓ be a line and let A and B be distinct points on the line. Then, by the **segment** (or line segment) with endpoints A and B, we mean the set of all points of the line lying between A and B including the points A and B (see Figure 5.1). The line ℓ is denoted by the symbol $\overleftrightarrow{AB}$ and the segment by the symbol $\overline{AB}$.

A concept more general than *line* is *curve*. By a **curve** (either in the plane or in space), we shall mean the trace of a moving point that moves in such a way that it does not skip any points. Examples are shown in Figure 5.2.

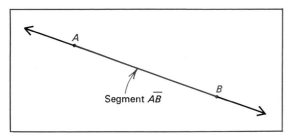

Figure 5.1

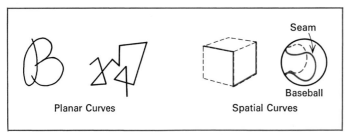

Planar Curves   Spatial Curves

Figure 5.2

A curve is called **closed** if it is the trace of a moving point that stops at the same point at which it started. A curve is called **simple** if it is the trace of a moving point that, between starting and stopping, does not pass through the same point twice (see Figure 5.3).

The most important kind of curve is a **simple closed curve.** A simple closed curve is both simple and closed; it is the trace of a moving point that stops at the same point at which it started and does not pass through the same point twice (see Figure 5.4).

The property of simple closed curves that makes them important is exceedingly intuitive and is displayed in the following important theorem, first discussed by Camille Jordan (French, 1838–1922).

The Jordan Curve Theorem.
Every simple closed curve in the plane separates the plane into three pairwise disjoint sets of points called the **interior** *of the curve, the curve itself, and the* **exterior** *of the curve* (see Figure 5.5).

By a **planar region** we shall mean a portion of the plane that is bounded by simple closed curves. The curves that are the boundaries of these regions are not themselves a part of the regions. The regions we shall be working with most either consist of the interiors of simple closed curves or will be annular. Refer to Figure 5.6.

It is difficult to give a careful definition of a **surface** in space, but examples of the more common surfaces are easy to find. Spheres, cubes, and cylinders are

Figure 5.3

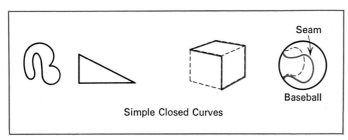

Figure 5.4

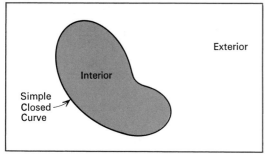

Figure 5.5

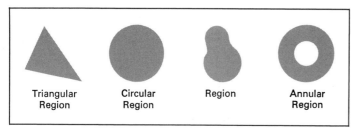

Figure 5.6

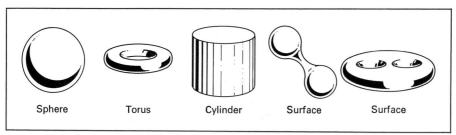

Figure 5.7

surfaces. The surface that corresponds to the annulus in the plane is called the **torus.** These and some other surfaces are shown in Figure 5.7. Each planar region is a surface when regarded as a spatial figure. The main thing to keep in mind is that surfaces are only "one point thick."

Exercises 5.1

1. Find examples in your environment of each of the following kinds of curves both in the plane and in space.
   (a) Curves that are neither simple nor closed.

(b) Curves that are simple but not closed.

(c) Curves that are closed but not simple.

(d) Simple closed curves.

2. Identify to which of the four classes mentioned in Exercise 1 each of these letters belongs: A, B, C, D, E, F, G, H, I, J, K, L, M, N, O, P, Q, R, S, T, U, V, W, X, Y, and Z.

3. Two lines in the plane must either intersect or be parallel. In space there is a third possibility. What is it?

4. Which of the following are true in space? If a statement is false, give an example of its falsity.

(a) Every two distinct lines are either parallel or intersect.

(b) Every two distinct planes that intersect, intersect in exactly one line.

(c) A line is uniquely determined by two distinct points.

(d) A plane is uniquely determined by two distinct points.

(e) Given a line and a point not on that line, there is exactly one plane containing both the line and the point.

(f) Given two distinct lines, they determine exactly one plane.

5. Imagine a point situated at the end of your nose. From the time you got up this morning until the time you go back to bed this evening that point will trace out a curve through space. Will this curve be simple? Will it be closed?

6. A bus running on its route might, under the right circumstances, trace out a simple closed curve. Can you think of any other simple closed curves traced out by the motion of familiar objects?

7. Give a number of examples of different kinds of surfaces found in your everyday environment.

8. Let $\mathcal{R}$ be a region in the plane bounded by a simple closed curve and let ℓ be a line. Is it possible that

(a) $\mathcal{R} \cap \ell$ might consist of a single point? (*Hint:* The curve that bounds the region is *not* a part of the region.)

(b) $\mathcal{R} \cap \ell$ might consist of a segment?

(c) $\mathcal{R} \cap \ell$ might consist of exactly two points?

(d) $\mathcal{R} \cap \ell$ might contain (as a subset) a segment?

(e) $\mathcal{R} \cap \ell$ might be the union of two disjoint sets?

9. Can you find examples of simple closed curves $\mathcal{C}_1$ and $\mathcal{C}_2$ in the plane such that

(a) $\mathcal{C}_1 \cap \mathcal{C}_2$ consists of just one point?

(b) $\mathcal{C}_1 \cup \mathcal{C}_2$ is a simple closed curve?

(c) $\mathcal{C}_1 \cup \mathcal{C}_2$ is a closed curve?

10. Let $\mathcal{C}$ be some simple closed curve in the plane, and let Y be a point lying in the exterior of this curve. Let X be point of the plane not belonging to the curve $\mathcal{C}$ itself. Draw the segment $\overline{XY}$. By counting the number of intersections this segment makes with the curve $\mathcal{C}$ you can determine whether X lies in the interior or in the exterior of the curve. How?

*11. We call a region in the plane **convex** if, given any two points of the region, the segment determined by them is contained entirely within the region. That is, $\mathcal{R}$ is convex if $\overline{XY} \subseteq \mathcal{R}$ for all points X and Y belonging to $\mathcal{R}$. Which of the regions in Figure 5.6 are convex? Which of the surfaces in Figure 5.7 enclose convex regions in space?

*12. Give an example of a convex region in the plane whose boundary is cut by every line in no more than two points. Can you find an example of a nonconvex region with the same property?

5.2 Congruence of Plane Figures

The objects of study in Euclidean geometry are geometric figures and, although we have identified only a few of these, we have enough to begin our study. There are two relations defined on geometric figures that are of particular importance in Euclidean geometry. We shall study the first of these in this section and the second in the next section.

The first relation defined on geometric figures that we shall discuss is the **congruence** relation. Loosely speaking, when we say that two figures are congruent we mean that it is possible to move one of the figures through space in such a way as to superimpose it upon the second figure. The pairs of figures in Figure 5.8 are congruent.

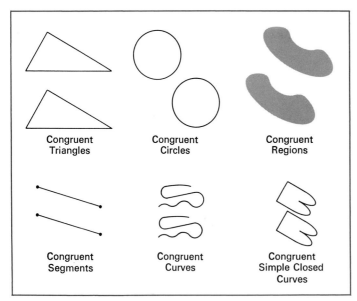

Figure 5.8

In the preceding rough description of what it means to say that two figures are congruent, we spoke of moving one figure through space in such a way as to superimpose it upon the second figure. We did not specifically say it, but you surely took it for granted that as the first figure is moved, it should not be bent, stretched, shrunk, torn, or otherwise deformed in any way. In other words, as we move the first figure we must keep it rigid. Hence the kind of motion we are talking about here is called **rigid motion.** A rigid motion is described by saying that such a motion of a geometric figure preserves distances between all pairs of points of that figure. Thus if points A and B are points of a certain geometric figure and are a certain distance apart, then, after that figure has been moved by a rigid motion, the points A and B are still that same distance apart. Here are some examples of rigid motions.

Example 1. The motion that moves a train along a straight track is a rigid motion. This particular kind of rigid motion is called a **translation.** When a sash window is opened or closed, the window is moved by means of a translation. The motion the carriage of a typewriter goes through as it moves along is also a translation.

Example 2. Imagine a ring binder lying flat open. When a page in the binder (see Figure 5.9) is turned, the page is moved through space with a rigid motion. This particular kind of rigid motion is called a **reflection.** The page is reflected about a line running down the spine of the binder. When a door is opened all

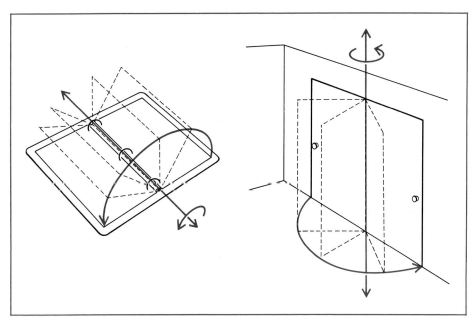

Figure 5.9

the way (180°), the door undergoes a rigid motion that is a reflection about a line running down the door jamb.

Example 3.   When a roulette wheel is spun, the wheel undergoes a rigid motion called a **rotation.** When a phonograph turntable is moved through a certain angle, the surface of the table is moved with a rotation. The motion here takes place in the plane of the table and the rotation is about the center of the turntable. The motion that moves the knob on a radio when stations are changed is a rotation.

In these examples we have described three distinct kinds of rigid motions: translations, reflections, and rotations. Translations and rotations do not take a planar figure out of the plane that contains it, but a reflection takes a figure out of its plane through space and then back into its plane again when the motion is completed. These three special rigid motions are important because of the following theorem.

Theorem.
Every rigid motion of a plane figure is a combination of one or more translations, reflections, or rotations.

These three special kinds of rigid motions are therefore the basic building blocks from which all rigid motions are constructed. Here are some examples.

Example 4.   Consider the two triangles shown in Figure 5.10. These two triangles are congruent because the left-hand triangle can be moved so as to coincide exactly with the right-hand triangle by means of rigid motions. In this case a translation alone will serve.

Example 5.   The triangles in Figure 5.11 are congruent; thus it should be possible to make the left-hand triangle coincide with the right-hand one. This can be accomplished by a combination of a translation and rotation, as shown in the

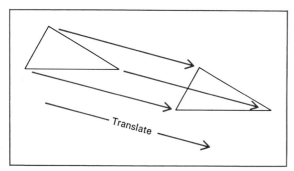

Translate

Figure 5.10

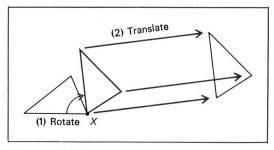

Figure 5.11

illustration. First, the left-hand triangle is rotated through a certain angle around the point X. Then the rotated triangle is translated until it coincides with the right-hand triangle.

Example 6. The two triangles shown in Figure 5.12 are congruent, and it is possible, therefore, to move the left-hand triangle using a rigid motion so that it will coincide with the right-hand triangle. We leave it to you to convince yourself that this cannot be accomplished without using at least one reflection. One way to accomplish this (there are many ways) is first to reflect the left-hand triangle about the line ℓ and then to translate the reflected triangle in the indicated direction until it coincides with the right-hand triangle.

The notion of congruence in space is very similar to that of congruence in the plane and can be discussed equally well in terms of translations, reflections, and rotations. However, we shall restrict our attention to the plane.

We shall return to the congruence relation in Section 5.4 after pausing to discuss a second relation on geometric figures.

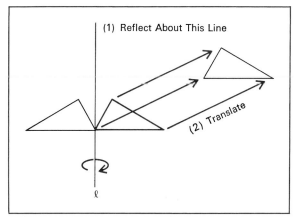

Figure 5.12

1. Give some everyday examples of motions that are translations, reflections, or rotations.
2. The following pairs of figures are congruent. Describe a rigid motion that will make the left-hand figure coincide with the right-hand figure.

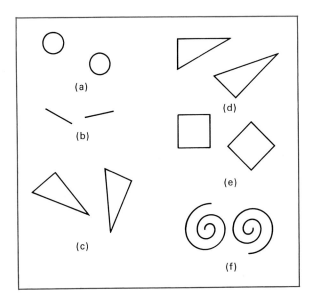

3. Draw two congruent triangles that can be brought into coincidence without using a reflection. Draw two that cannot be brought into coincidence without using a reflection.
4. Find some examples of rigid motions that you use in everyday life to make congruent figures coincide. (For example, stacking plates.)
5. Explain, using rigid motions, why the relation *is congruent to* is an equivalence relation on plane figures.
6. A triangle was moved by a rigid motion in such a way that exactly one point of the triangle was left unmoved—that is, was left fixed by the motion. What kind of a motion was this?
7. A triangle was moved by a rigid motion. In the process at least one point of the triangle was moved to a different position in the plane but two (or possibly more than two) points of the triangle were not moved. What can you say about the motion?
8. A triangle was moved by a rigid motion in such a way that three noncollinear points were left fixed by the motion. What can you say about the motion?
9. What kinds of rigid motions are needed to make two congruent circles coincide? To make two congruent squares coincide? To make two congruent rectangles coincide? To make two congruent isosceles triangles coincide?

5.3 Similarity of Plane Figures

The second relation on geometric figures that is important in Euclidean geometry is the similarity relation. This relation is a generalization of the congruence relation in the sense that any two congruent figures are also similar.

An imprecise but partially meaningful definition of congruence is that two geometric figures are congruent if and only if they have the same shape and the same size. (The trouble with this statement is that the concepts of "size" and "shape" have not been carefully defined.) In this same vein, we might say that two geometric figures are *similar* if they have the same shape but not necessarily the same size. The pairs of figures in Figure 5.13 are similar but not congruent. Our purpose in this section will be to formulate a more precise definition of similarity in terms of motions.

It is obvious that similar figures cannot be made to coincide using just the rigid motions. What is needed is a new kind of motion which, together with the rigid motions, can be used to make similar figures coincide. Clearly, this new motion must be one that stretches or shrinks figures. This new kind of motion is called a "homothety."

A **homothety** is a motion that has the effect of moving the individual points of a geometric figure inward toward, or outward from, a fixed point called the **center** of the homothety. The motion of the points with respect to this center is *uniform* in the following sense. Suppose (see Figure 5.14) that X and Y are points of a certain geometric figure and that this figure is to be moved by a homothety whose center is at the point C. If, before the homothety is applied to the figure, the distances of X and Y from the center C are x and y, respectively, and if the distance between X and Y is d, then, after the homothety has been applied, the distances of X and Y from C are rx and ry, respectively, and the distance between X and Y is rd, where r is some number. This number r is called the **ratio** of the homothety.

A magnifying glass accomplishes a homothety that enlarges figures. It moves points outward away from the central point P, which lies directly under the center

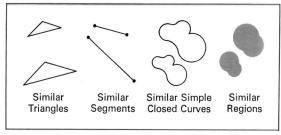

Similar Triangles Similar Segments Similar Simple Closed Curves Similar Regions

Figure 5.13

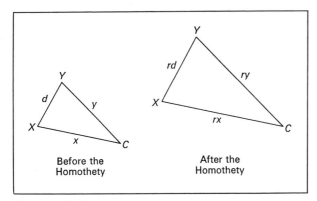

Figure 5.14

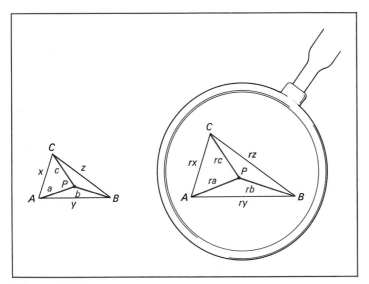

Figure 5.15

of the glass (refer to Figure 5.15). The ratio of the distances of the points from this central point, when viewed through the glass, to their distances from this central point, when viewed without the glass, is a measure of the magnifying power of the glass. The ratio of the homothety displayed in Figure 5.15 is r.

Now we can say what it means for two geometric figures to be similar. Two figures are **similar** if it is possible to make one of the figures coincide exactly with the other figure by moving it by means of a homothety and one or more of the rigid motions discussed in the preceding section. Here are some examples of similar figures and the motions that can be used to make one of the figures coincide with the other.

Example 1. The circle $\mathcal{C}_1$ shown in Figure 5.16 can be moved so that it will coincide with the circle $\mathcal{C}_2$ as follows. First, apply a homothety whose center is at the center X of circle $\mathcal{C}_1$ and whose ratio is $5:3$. This homothety will produce an enlarged circle of radius 5 and center at X. Then a translation will move this enlarged circle to coincide with circle $\mathcal{C}_2$. Hence the circles $\mathcal{C}_1$ and $\mathcal{C}_2$ are similar.

Example 2. The two triangles shown in Figure 5.17 are similar. It is possible, therefore, to make the left-hand triangle coincide with the right-hand triangle by applying an appropriate combination of a homothety and some rigid motions. We have first applied a homothety with ratio $1:2$ and center at the point P. The resulting triangle is congruent to the right-hand triangle so all that we need to do is to rotate and translate the shrunken triangle to make it coincide with the right-hand triangle.

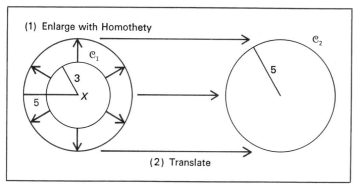

Figure 5.16

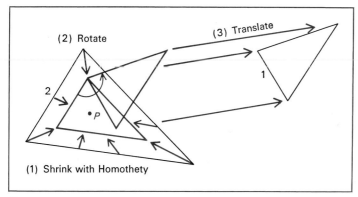

Figure 5.17

Applications of similarity are easy to find. We have already mentioned the magnifying glass. Scale drawings of all kinds, model airplanes and trains, dolls and toy soldiers, all utilize this relation.

The notion of similarity applies equally well for spatial figures as for planar figures. However, we shall deal only with similarity in the plane.

Exercises 5.3

1. Find some examples of motions experienced in everyday life that at least resemble homotheties.
2. The following pairs of figures are similar. Describe rigid motions and homotheties that can be used to make the left-hand figure coincide with the right-hand figure.

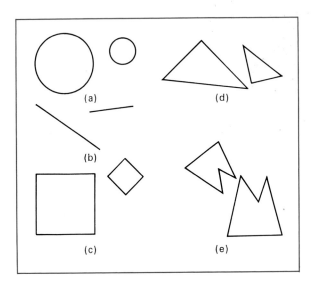

3. If the ratio of a homothety is greater than 1, in what way does the homothety change the figure? What happens to the figure if the ratio is less than 1? What happens if the ratio is equal to 1?
4. How is the idea of a homothety involved in maps? How is it involved in model airplanes?
5. Construct a square of side 1 inch. Now move that square by applying a homothety of ratio 2 and center at one of the vertices of the square. Now move the square again, this time with a homothety of ratio 2 and center at the center of the square. How do the two new squares obtained by moving the given square with these two homotheties compare with each other? Can you draw a generalization from this experiment?

6. When moved by a certain homothety, two points of a figure were left unmoved. What is the ratio of the homothety? What was the effect of the homothety upon the other points of the figure?

7. Explain in terms of rigid motions and homotheties why the relation *is similar to* defined on the set of all plane figures is an equivalence relation.

*8. If ℜ is a convex region and if ℜ is moved by a homothety, is the resulting region also convex?

9. If a square is moved by a homothety of ratio r, how does the area of the original square compare with the area of the square that results from the homothety motion?

5.4 What Is a Geometry?

As we observed in Chapter 1, by the middle of the nineteenth century there were a number of different geometries being studied, among which were Euclidean geometry and the non-Euclidean geometries of Lobachevsky and Riemann. Each of these geometries was defined in terms of the axiom system upon which it was built. Then, in 1872 Felix Klein introduced a method for classifying geometries, not according to their axiom systems but instead in terms of what are called "invariant properties." In this section we shall describe Klein's codification of geometries and use this method to define Euclidean geometry. In the next section we shall use Klein's method to introduce another geometry called "topology."

Suppose that $\mathcal{F}$ is a geometric figure and that we move $\mathcal{F}$ by using one or more of the rigid motions so as to obtain a second figure $\mathcal{F}'$. If $\mathcal{F}$ has the property of being a circle, what can you say about $\mathcal{F}'$? Surely, $\mathcal{F}'$ is also a circle. We say that the property of being a circle is preserved by rigid motions and call this property a congruence invariant. **A congruence invariant** is a property of geometric figures that is preserved when those figures are moved by rigid motions. Some other congruence invariants are

1. The property of being a triangle, square, circle, and so forth.
2. If two lines make a certain angle with each other, then, when this configuration of lines is moved by a rigid motion, the angle between the lines is unchanged. Hence angles are preserved by rigid motions and so *angle* is a congruence invariant.
3. If a region has a certain area and is moved by a rigid motion, then the resulting region has the same area. Hence *area* is a congruence invariant.
4. If two points are situated a certain distance apart, then, after these points are moved by using a rigid motion, they remain the same distance apart. Hence *distance* is a congruence invariant.

A **similarity invariant** is a property preserved by the motions that define the relation of similarity—homotheties and rigid motions. The properties of being a triangle, square, circle, and so on, are similarity invariants. *Angle* is also a similarity invariant. However, *area* and *distance* are not similarity invariants because neither the area of regions nor the lengths of segments are preserved by homotheties. Every similarity invariant is also a congruence invariant (because properties preserved by homotheties and rigid motions are preserved by rigid motions alone), but not every congruence invariant is a similarity invariant.

The importance of these invariants was first discussed by Felix Klein (German, 1849–1929) in 1872. Klein defined Euclidean geometry to be the mathematical study of congruence and similarity invariants. That is to say, Euclidean geometry is the study of geometric properties that are preserved by rigid motions or by homotheties and rigid motions. Actually, Klein was considerably more general that this. There are a wide variety of different geometries, Euclidean geometry being only one of them.

Klein's codification of geometries according to their invariants goes something like this: Suppose that $\Re$ is a relation on geometric figures.[1] Then it is possible to define this relation in terms of certain kinds of motions. (For example, the relation of congruence is defined in terms of rigid motions.) The properties of figures that are preserved by these motions are called the invariants for the relation $\Re$. The study of these invariants is called $\Re$-geometry.

As an example of how Klein's method of describing a geometry works, consider the (equivalence) relation of congruence. This relation can be defined in terms of the rigid motions alone. The invariants of these rigid motions are the congruence invariants. The study of these congruence invariants is called **congruence geometry.** Congruence geometry is different from Euclidean geometry because Euclidean geometry utilizes the similarity relation as well as the congruence relation.

As another example of a geometry, we may consider **similarity geometry.** We begin with the relation of similarity and then define this relation in terms of motions. The properties left unchanged by these motions are the similarity invariants; similarity geometry is the study of these invariants.

To illustrate the differences between congruence geometry, similarity geometry, and Euclidean geometry, consider the two figures in Figure 5.18. The congruence geometer would look at this illustration and see two completely different figures. This is all he could say about these figures—they are different because they are not congruent. The similarity geometer would look at these figures and tell you that they are the same. He could not distinguish between them because they are similar. The Euclidean geometer would look at these (just as you do) and say that, although they are in one sense different figures, in another sense they are the same too. That is, the Euclidean geometer could see differences and similarities between these figures.

In the presence of Euclidean geometry, congruence geometry and similarity

---

[1]For technical reasons the relation $\Re$ must be an equivalence relation, but since we are not going to get technical we may ignore this.

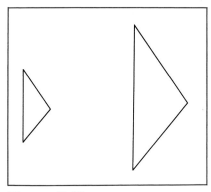

Figure 5.18

geometry are usually ignored. We have discussed them only because we wanted to show by means of very simple examples how Klein's codification of geometries works. There are a number of other geometries, however, that are very unlike Euclidean geometry and have been very important in the historical development of mathematics. Of these we wish to discuss only the one that is far and away the most important, the geometry called *topology*.[2] We shall do this in the next two sections.

Exercises 5.4

1. Which of the following properties are congruence invariants? Which are similarity invariants?
   (a) The property of being two parallel lines.
   (b) The property (of simple closed curves) of containing an area of 10 square inches.
   (c) The property of being an equilateral triangle.
   (d) The property of being a simple closed curve.
   (e) The property of being a closed curve that is not simple.
   (f) The property of being two perpendicular lines.
   (g) The property (of rectangles) of having a length to width ratio of 2 to 1.
   (h) The property of being intersecting lines.
   (i) The property of being a segment.
   (j) The property of being a segment 2 inches long.
*2. Is the property of being a convex region a congruence invariant? Is it a similarity invariant?
3. Is every congruence invariant a similarity invariant? Is every similarity invariant a congruence invariant?

[2] *Topology* comes from the Greek words *topo* (space) and *logia* (study).

4. If a congruence geometer were to look at the figures shown below, how many different figures would he see? How many different figures would a similarity geometer see?

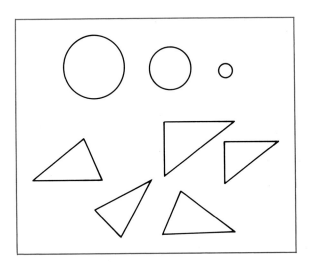

5.5 Topological Equivalence

Of the geometries that are substantially different from classical Euclidean geometry, topology is by any measure the most important. Topology arrived on the scene about 20 years after Cantor's set theory and evolved from the work of a number of different mathematicians working in different parts of mathematics. Henri Poincaré (French, 1854–1912) is regarded as the principal developer of modern topology. Along with set theory, to which it is closely allied, topology has invaded all parts of mathematics and is one of the basic subjects upon which modern mathematics depends.

We are going to arrange our introduction to topology along the lines of Klein's definition of a geometry. We begin by introducing a new (equivalence) relation and the motions that define it. Then we shall look at a few of the simpler invariants for that relation. Topology is the study of these invariants.

Place a rubber band on a table and, using your fingers, move the rubber band around so as to deform it in various ways. For example, the rubber band can be moved into the shapes shown in Figure 5.19. Such deforming motions are called **deformations.** Roughly speaking, a deformation of a figure is a motion that changes the shape and the size of a figure in ways subject only to the following three restrictions: (1) A deformation cannot puncture a figure. That is, it cannot remove

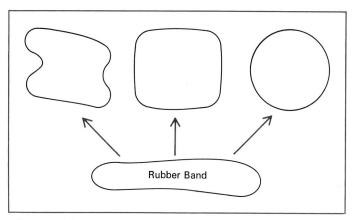

Figure 5.19

a point from a figure. (2) A deformation cannot tear a figure.[3] For example, it cannot deform a circle into two separate circles. (3) A deformation cannot amalgamate points. That is, it cannot bring distinct points together into one point. For example, it cannot deform a line segment in such a way that it becomes a single point.

Here are some examples of deformations of planar figures.

Example 1. The left-hand region in Figure 5.20 can be deformed in such a way that, after it has been deformed, it will be congruent to the right-hand region. The deformation that is required is one that will shrink the "ends" of the region and stretch the "middle."

Example 2. By appropriately deforming the figure on the left in Figure 5.21, it can be made congruent to the figure on the right. This is accomplished by deforming the circle into a triangle, stretching and bending the short vertical segment, and shrinking the longer horizontal segment.

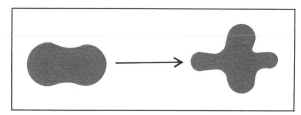

Figure 5.20

[3]It is permissible to tear a figure, deform it, and then repair the tear so that the edges fit together exactly as they did before the tear was made. We shall, however, not become involved with such deformations in our work.

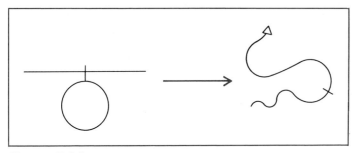

Figure 5.21

Example 3. In Figure 5.22, figure 𝔞 can be deformed so as to be congruent to figure 𝔅, but cannot be deformed (without amalgamating points) so as to be congruent to figure 𝔠.

When two figures can be made to coincide using rigid motions, they are called congruent. When they can be made to coincide using homotheties and rigid motions, they are called similar. When they can be made to coincide using deformations and rigid motions, they are called **topologically equivalent.** Here are more examples of topologically equivalent figures.

Example 4. The pairs of figures in Figure 5.23 are topologically equivalent.

Example 5. The pairs of figures in Figure 5.24 are not topologically equivalent. In each case the left-hand figure could be deformed into the right-hand figure only by puncturing the figure, tearing the figure, or amalgamating some points of the figure.

If a congruence geometer is a person who cannot tell the difference between two circles of 1-inch radius, and if a similarity geometer is a person who cannot tell the difference between a circle of 1-inch radius and a circle of 2-inch radius, then a topologist is a person who cannot tell the difference between a circle and a square. This is because a circle and a square are topologically equivalent. A

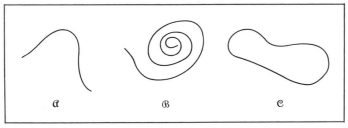

Figure 5.22

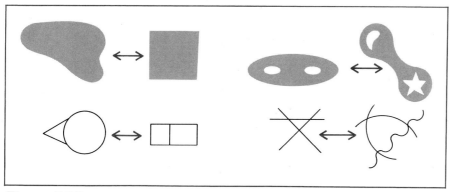

Figure 5.23

topologist can tell the difference between two geometric figures only if those figures are not topologically equivalent. Thus a topologist should not even be able to read, because he cannot tell the difference between the letters C, L, M, N, S, U, V, W, and Z. These letters are all topologically equivalent and should look the same to the topologist.

Now we can list some of the invariants of the relation of topological equivalence. Here is a partial list of such invariant properties.

1. *The property of being a simple closed curve.* (A curve is a simple closed curve if and only if it is topologically equivalent to the circle.)
2. *The property of being* **simply connected.** A region (in the plane) is called simply connected if every simple closed curve contained within the region can be shrunk to a point without leaving the region. In Figure 5.25 the region 𝔞 is simply connected, whereas region 𝔅 is not.

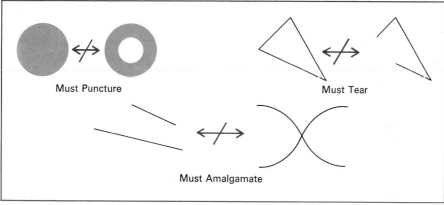

Figure 5.24

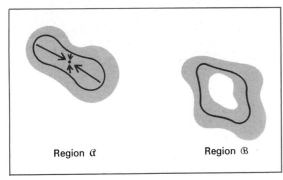

Figure 5.25

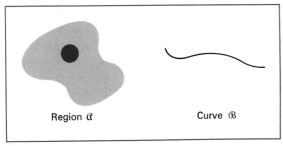

Figure 5.26

3. *The property of containing a circular region.* In Figure 5.26 region 𝕮 contains a circular region while curve 𝕭 does not. Hence region 𝕮 is not topologically equivalent to curve 𝕭.

4. *The property of consisting of a finite set of points.* Since a deformation cannot amalgamate points, if two figures are topologically equivalent, then each figure must contain the same (cardinal) number of points. Hence, if one figure is finite, then so is the other.

5. *The property of being* **arcwise connected.** A planar region is called arcwise connected if it is possible to join every point of the region to every other point of the region by an arc (that is, a curve) that is contained entirely inside the region. The interior of a circle is arcwise connected, but the interior of a figure eight curve is not.

Example 6. Which of these figures in the plane are simply connected? Which contain circular regions? Which are arcwise connected?

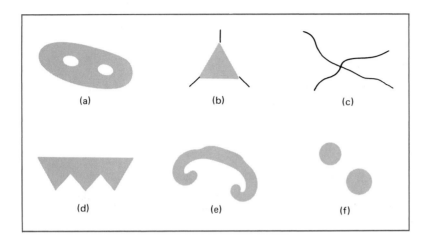

Solution: Figures (b), (c), (d), (e), and (f) are simply connected. Figures (a), (b), (d), (e), and (f) contain circular regions. Figures (a), (b), (c), (d), and (e) are arcwise connected.

Topology deals with only the most general properties of geometric figures in contrast to Euclidean geometry, which becomes deeply enmeshed with rather precise properties of figures. For example, in Euclidean geometry we differentiate between equilateral triangles, isosceles triangles, right triangles, acute triangles, obtuse triangles, and so on. But in topology these triangles are indistinguishable from one another. More than this, in topology we do not even distinguish between triangles and circles. Hence a standard theorem from Euclidean geometry such as, "the diagonals of a rectangle bisect each other" or "the sum of the angles of a triangle is 180°" is meaningless in topology. In the study of topology, there is no such figure as a rectangle or a triangle; the nearest we can get to these figures is a simple closed curve. A theorem from topology, on the other hand, will also make sense in Euclidean geometry. An example is the Jordan Curve Theorem mentioned in Section 5.1. This theorem is really a topological theorem, because it deals only with ideas that are meaningful in topology: simple closed curves, interiors, and exteriors.

In the next section we shall discuss two of the more interesting surfaces that are important in topology.

Exercises 5.5

1. Give some examples of deformations that are used in everyday life. (For example, crumpling up a piece of paper.)
2. Is a rigid motion a topological deformation? Is a homothety a topological deformation?

3. Find topologically equivalent figures among the following figures.

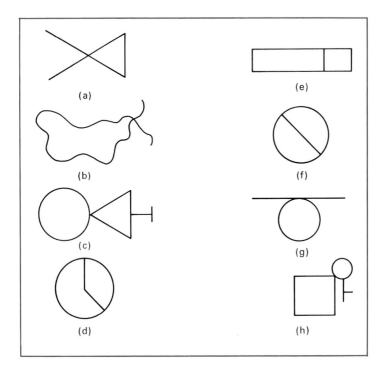

(a)

(e)

(b)

(f)

(c)

(d)

(g)

(h)

4. Explain why the relation *is topologically equivalent to* is an equivalence relation on the collection of all figures.

5. Draw pictures of three curves topologically equivalent to a figure eight. To a circle. To the block letter A. Describe a surface that is topologically equivalent to the surface of a doughnut.

6. Which of the figures on p. 151 are simple closed curves? Which are simply connected? Which contain a circular region? Which consist of a finite set of points? Which are arcwise connected?

7. Separate the letters of the alphabet (when written as simple block letters) into pairwise disjoint sets such that no letter from one set is topologically equivalent to any letter of any different set, but any two letters in the same set are topologically equivalent. Use the letters in Exercise 2, Section 5.1.

8. Which of the properties listed in Exercise 1, Section 5.4, are topological invariants?

*9. Is the property of being a convex region a topological invariant? Is a convex region always simply connected? Is a simply connected region necessarily convex? Is a convex region necessarily arcwise connected? Is every arcwise connected region convex?

10. Is every arcwise connected region simply connected? Give an example.

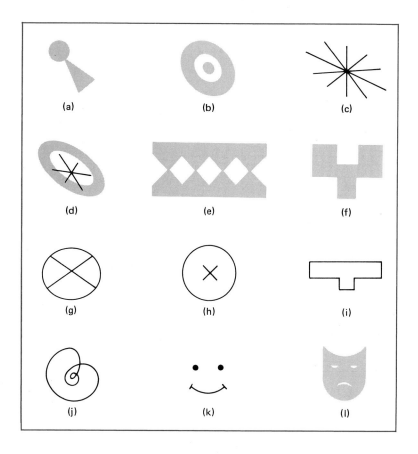

(a) (b) (c) (d) (e) (f) (g) (h) (i) (j) (k) (l)

11. If a topologist is a person who can't tell the difference between two equivalent figures, then why can't a topologist tell the difference between a doughnut and a coffee cup?

12. Is the interior of a simple closed curve simply connected? Is it arcwise connected? Is the exterior simply connected? Is it arcwise connected?

*13. A problem of great historical interest to topology is the problem of the **Konigsberg bridges.** (Konigsberg is now called Kaliningrad and lies in East Germany.) Through the city ran two branches of a river at whose confluence there was an island. Seven bridges connected the island to the banks. The problem was to try to find a path that would cross each bridge once and only once. Is there such a path? Refer to the figure on p. 152.

*14. Imagine that each country is a connected land area (not like Michigan or Hawaii) and that we are to make a map using the least possible number of colors. If two countries share a border consisting of a curve, then they must be colored differently. However, if they only share a border consisting of a point (like Arizona and Colorado), then they may be colored the same. The **Four-Color Problem** asks whether every map can be made with, at most, four

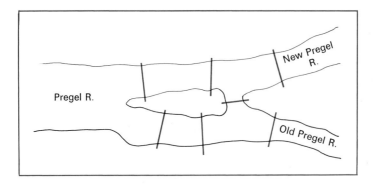

colors. (This can be any imaginary map—it does not have to be a map of the real world.) There are maps that require four colors and no one has ever found a map that required five colors, but neither has anyone been able to prove that only four colors would ever be needed. This is one of the famous unsolved problems of topology and has been under study since about 1840. Show that any map of the United States would require four colors. (*Hint:* Consider the states in the vicinity of either Nevada or Kentucky and see that four colors would be required to color them.)

15. Some states like Nevada and Colorado are arcwise connected.
    (a) Name some states that are arcwise connected and some that are not.
    (b) Are there any states that are not simply connected?
    (c) Can you think of any countries in the world that are not simply connected? That are not arcwise connected?
    (d) Are there any states that do not contain circular regions?

5.6 One-Sided Surfaces

Compared with the kinds of geometric figures dealt with in topology, most of the figures of importance in Euclidean geometry are prosaic indeed. In this section we shall discuss two "topological" surfaces, each of which has only one side. These one-sided surfaces were not even invented until mathematicians began to study topology, there being no motivation to study such exotic figures in Euclidean geometry. Euclidean geometry, after all, derives from the need to find a mathematical representation for the useful physical objects we experience in our real world. Topology is not restricted to this "practical" sort of figure but may deal with any sort of figures, whether they have a practical existence in our experience or not.

Make two pieces of paper about 2 inches wide and 24 inches long and label the vertices as shown in Figure 5.27(a). If these vertices are joined as shown in Figure 5.27(b) and then taped together, the surface that results is called a **right circular cylinder.** This is a very familiar surface, which is of considerable importance in Euclidean geometry. This surface has two sides and two edges, as shown in the figure. In order to move from one side to the other, it is necessary to cross over one of the edges.

In 1858, a German, A. F. Moebius (1790–1868), varied this construction by taking a half-twist in the strip before taping the ends together. He joined the four vertices, as shown in Figure 5.27(c). The surface that results is called the **Moebius strip** and has some unusual properties. First, this strip has no inside and no outside—it has only one side. To show this, select any point on the surface of the strip. With a pencil, start to move along this surface, as shown in Figure 5.28. You will find that in doing this you traverse the entire strip and come back to the point at which you started. While traversing the surface in this way, you do not cross over an edge, and you must conclude that the surface is one-sided. Second, this surface has only one edge. Start at any point on the edge and traverse along the edge. You will arrive back at your starting place and in the meantime you will have traversed the entire edge of the strip.

The Moebius strip and the right circular cylinder are not topologically equivalent. Both the number of sides of a figure and the number of edges of a figure are topological invariants. Since the cylinder and the strip do not have the same

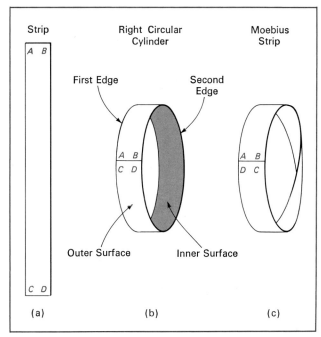

Figure 5.27

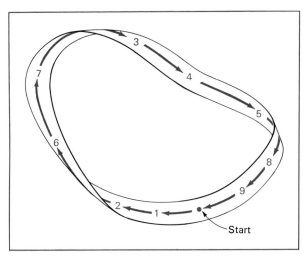

Figure 5.28

number of sides (or the same number of edges), these surfaces cannot be topologically the same.

The **punctured Klein Bottle** was invented by Felix Klein in 1882. It is possible to make a physical model of the punctured Klein bottle, but this would require a certain amount of talent. We must be content with planar pictures of this bottle, which are necessarily somewhat difficult to visualize. The drawings in Figure 5.29 demonstrate how to make a punctured Klein bottle. Begin with an ordinary bottle that is made out of a completely elastic glass. Then make a hump in the bottom and cut off the top of this hump. Also cut a hole in the side of the bottle. Now stretch out the neck of the bottle and pull it around and pass it through the hole you cut in the side of the bottle. Join the mouth of the bottle neatly to the edge

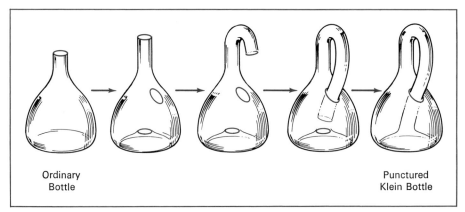

Ordinary
Bottle

Punctured
Klein Bottle

Figure 5.29

formed when you cut the hole in the bottom of the bottle. The resulting surface is called the punctured Klein bottle.

The punctured Klein bottle has only one edge, the edge formed when you cut the hole in the side of the original bottle. Also, the bottle has no outside or inside—it is a one-sided surface. To see this, study the illustration until you are convinced that you can trace from any one point on the surface to any other point on the surface without crossing over the edge of the surface.

The punctured Klein bottle is not topologically equivalent to the ordinary bottle because the ordinary bottle has two sides and the punctured Klein bottle has only one side. As both the Moebius strip and the punctured Klein bottle have one edge and one side, on the basis of these two invariants alone we cannot distinguish between these two figures. In order to show that these two figures are not topologically the same, we shall introduce another invariant.

By a **loop** on a surface we mean a simple closed curve on the surface that does not cross over an edge of the surface. Examples of loops are shown in Figure 5.30.

The largest number of loops that can be cut in a *one-edged surface* without separating that surface into two or more pieces is a topological invariant. Let us call this number the **loop number** of a one-edged surface. For example, the disk is one-edged and has loop number 0, since any one loop cut will separate the disk into two pieces. The punctured torus (like an inner tube with the valve stem removed) is a one-edged surface with loop number 1. The reason for this is that any two loop cuts will separate the punctured torus into two separate pieces, but it is possible to cut one loop in the figure without separating it into two pieces.

Using your paper Moebius strip, you can see for yourself that, if you make a loop cut that runs right down the middle of the strip, the strip is not separated into two pieces. The surface that results is shown in Figure 5.31. It follows that the loop number of the Moebius strip is at least 1. But, as soon as you make another loop cut, the surface will separate into two pieces. You can experiment and see that this is true. Hence the loop number of the strip is 1.

If you do not have a model of the punctured Klein bottle to work with, it is

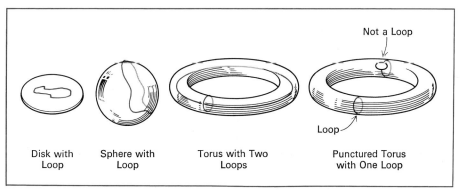

Disk with Loop   Sphere with Loop   Torus with Two Loops   Punctured Torus with One Loop

Figure 5.30

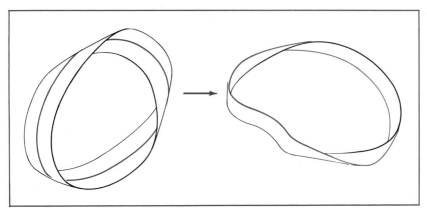

Figure 5.31

harder for you to see that the loop number of the punctured Klein bottle is 2, but such is the case. For example, the two loop cuts shown in Figure 5.32 do not separate the punctured Klein bottle, but any three loop cuts will separate the bottle into two pieces. Hence the loop number is 2.

Since the strip and the bottle have different loop numbers, these surfaces are not topologically equivalent.

We have got a glimpse of just a few topological invariants in this and the

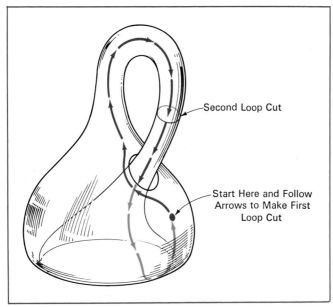

Second Loop Cut

Start Here and Follow Arrows to Make First Loop Cut

Figure 5.32

preceding section. The last one, the loop number, illustrates how far removed topology is from Euclidean geometry. Of these two geometries, topology is far and away the more interesting and useful to the modern mathematician. Topology is the object of intense study by researchers, whereas almost no one does research in Euclidean geometry any more. Euclidean geometry, like Latin, is not thriving, but it is always around us and is always in use. Topology is more like a living language, constantly changing and expanding to satisfy new needs.

Exercises 5.6

1. The Moebius strip has one side and one edge. How many sides and edges do the spatial surfaces shown have?

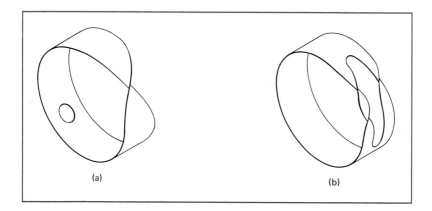

(a)

(b)

2. Is it possible to cut a loop in the Moebius strip that separates the figure into two separate pieces. Find such a loop cut. Why doesn't this mean that the loop number of the Moebius strip is 0? (*Hint:* Don't cut the loop down the middle.)
3. Find two loop cuts that separate the punctured Klein bottle into two separate pieces. Why doesn't this mean that the loop number of this figure is 1?
4. If you were to rip this page out of this book, you would be holding a one-edged surface in your hand. What would be the loop number of that surface?

   In order to answer the following exercises, you will need to make models. Strips of paper about 2 inches wide and 2 feet long would be best.

5. You can make a Moebius strip from a strip of paper by taking a half-turn in it before taping the ends together. Suppose that you take two strips of paper laid together, and keeping these pieces together, take a half-turn in them and then tape the pairs of ends together. You get what looks like two Moebius strips nestled together. Now, while keeping these strips together, observe that you can

run your finger around between them with no obstructions. Release the figure and see what you have. Compare what you have with the figure on the right-hand side of Figure 5.31. How many edges and sides does this figure have? This figure can be made from a strip of paper by taking how many half-turns in the strip before taping?

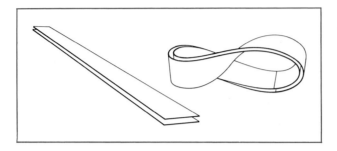

6. Take a strip of paper and, instead of taking one half-turn in it before taping (this would give you a Moebius strip), take one full turn before taping. How many edges and sides does this figure have? Then make a loop cut down the middle of this figure. What happens? How do the strips that result from this loop cut compare to the original strip?

7. Take a strip of paper and make one and one-half turns in it before taping the ends together. How many edges and sides does this figure have? Make a loop cut down the center of the strip and see what happens. Surprise! How many edges and sides does the resulting figure have? What is the loop number of the strip that you made?

8. We saw that making a loop cut down the middle of a Moebius strip did not separate the figure. Make a loop cut about one third of the way from one edge of the Moebius strip and see how the strip separates. Analyze each of the two strips so obtained. How many turns does each contain?

6

Extending the Concept of Number

In this chapter we shall continue from Chapter 3 the discussion of the notion of number. We shall carry the development of the number concept from its beginnings in the system of whole numbers to its first-stage culmination in the system of real numbers. Our concern in this chapter will be only partly with the usual computational aspects of working with numbers. Much of our attention will be directed toward the number systems themselves rather than just to the numbers that are the objects of the systems. By so doing, we shall be introducing some of the most basic ideas of that part of modern mathematics called "abstract algebra."

6.1 The System of Integers

Something about the idea of negative numbers makes them hard to accept as being legitimate numbers—at least this is the lesson taught us by history. In 274 A.D. the equation $4x + 20 = 4$ was called absurd (it has a negative solution). Negative numbers were called *numeri absurdi* by at least one important German mathematician of the early sixteenth century. As late as the end of the sixteenth century, some mathematicians, when discussing solutions of equations, studiously ignored the existence of solutions that were negative numbers. Descartes (of whom we shall have more to say in Chapter 7) referred to positive solutions of equations as "true" solutions and to negative solutions as "false" solutions. By the eighteenth century, however, negative numbers were accepted, but they were still not well understood. For instance, some textbooks of the period taught that it was impossible to multiply two negative numbers together. Even in the nineteenth century we can find many instances of negative numbers which, although understood, were simply ignored for no good mathematical reason. The whole history of these numbers seems to be one of rejection until the middle of the nineteenth century, when new developments in mathematics made it finally and abundantly clear that the negative numbers were as necessary to mathematics as the familiar positive numbers. In this section we shall introduce some negative numbers into the system of whole numbers and, as a result, produce the number system called "the system of integers."

As our first step toward enlarging the number system, we shall augment the set of whole numbers with the numbers that are called **negative integers.** These are the numbers whose most familiar names are -1, -2, -3, -4, and so on. The negative integer -1 is defined to be that number which gives 0 when 1 is added to it; that is, -1 is the solution of the equation $x + 1 = 0$. The negative integer -2 is the solution of the equation $x + 2 = 0$; -3 is the solution of the equation $x + 3 = 0$; and so on. For consistency of terminology we shall rename the nonzero whole numbers and call them **positive integers.** The number 0 is an integer but is neither positive nor negative. By an **integer** we mean a number that is zero, a positive integer, or a negative integer.

We can identify the integers with points on the number line. We begin by selecting two distinct points, which we call the **origin** and the **unit,** with which we associate the integers 0 and 1. Then we locate the other integers on the number line by marking off unit distances to the right and to the left of the origin (see Figure 6.1).

Like the system of whole numbers, the system of integers is equipped with two binary operations called "addition of integers" and "multiplication of integers." These operations, defined on the set of integers, are extensions of the corresponding operations defined on the set of whole numbers. That is, the result of applying addition of integers (or multiplication of integers) to two nonnegative integers (that is, whole numbers) is the same as the result of applying addition (or multiplication)

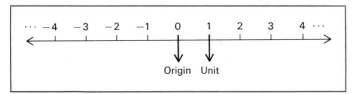

Figure 6.1

of whole numbers to those two numbers. These binary operations possess the following basic properties.

1. Addition of integers:
   (A) Is commutative: $x + y = y + x$.
   (B) Is associative: $x + (y + z) = (x + y) + z$.
   (C) Has the identity 0: $x + 0 = 0 + x = x$.
2. Multiplication of integers:
   (A) Is commutative: $x \cdot y = y \cdot x$.
   (B) Is associative: $x \cdot (y \cdot z) = (x \cdot y) \cdot z$.
   (C) Has the identity 1: $x \cdot 1 = 1 \cdot x = x$.
3. Addition and multiplication of integers share the left and right distributive properties:
   (A) $x \cdot (y + z) = (x \cdot y) + (x \cdot z)$.
   (B) $(x + y) \cdot z = (x \cdot z) + (y \cdot z)$.

In the system of whole numbers, addition and multiplication possess these same properties. The difference between these two number systems is that addition in the system of integers possesses one additional property of great importance.

1D Each integer possesses exactly one inverse with respect to addition.

This means that corresponding to each integer x there is another integer y, called the **additive inverse** of x, such that $x + y = y + x = 0$. For example, the additive inverse of 6 is -6, and the additive inverse of -5 is 5. Zero is its own additive inverse. If we denote the additive inverse of an integer x by the symbol $-(x)$, then we can render these statements symbolically as $-(6) = -6$, $-(-5) = 5$, and $-(0) = 0$. We shall see that it is this property 1D that makes the system of integers much more useful than the system of whole numbers.

Let us demonstrate by means of a few examples how these properties are used to verify addition and multiplication facts in the system of integers.

Example 1.   By using the stated properties of addition in the system of integers, explain why $-3 + -4 = -7$.

Solution: The sum of -3 and -4 must equal -7 because, when we add 7 to $-3 + -4$, the result is 0 and -7 is the only number with this property.

Observe:

$$
\begin{aligned}
(-3 + -4) + 7 &= (-3 + -4) + (4 + 3) & &(7 = 4 + 3) \\
&= -3 + [-4 + (4 + 3)] & &(1B) \\
&= -3 + [(-4 + 4) + 3] & &(1B) \\
&= -3 + [0 + 3] & &(-4 \text{ and } 4 \text{ are} \\
& & &\text{additive inverses})
\end{aligned}
$$

$$= -3 + 3 \qquad \text{(1C)}$$
$$= 0. \qquad (-3 \text{ and } 3 \text{ are}$$
$$\text{additive inverses)}$$

Example 2. Explain why $-3 + 5 = 2$.

Solution: We write $5 = 3 + 2$ and proceed as shown:

$$\begin{aligned} -3 + 5 &= -3 + (3 + 2) &\quad& (5 = 3 + 2) \\ &= (-3 + 3) + 2 &\quad& \text{(1B)} \\ &= 0 + 2 &\quad& (-3 \text{ and } 3 \text{ are additive inverses)} \\ &= 2. &\quad& \text{(1C)} \end{aligned}$$

Example 3. Explain why $-7 + 4 = -3$.

Solution: We write $-7 = -3 + -4$ (see Example 1) and proceed as follows:

$$\begin{aligned} -7 + 4 &= (-3 + -4) + 4 &\quad& \text{(From Example 1)} \\ &= -3 + (-4 + 4) &\quad& \text{(1B)} \\ &= -3 + 0 &\quad& (-4 \text{ and } 4 \text{ are additive inverses)} \\ &= -3. &\quad& \text{(1C)} \end{aligned}$$

The flow chart in Figure 6.2 contains a procedure that can be used to add any two integers.

Example 4. We already know that 0 is an annihilator for whole numbers. Zero is also an annihilator for the negative integers. For example, consider the product of -3 and 0. We argue as follows:

$$\begin{aligned} (-3)(0) &= (-3)(0) + 0 \\ &= (-3)(0) + (3)(0) \\ &= (-3 + 3)(0) \\ &= (0)(0) \\ &= 0. \end{aligned}$$

You should justify each of the steps in this argument.

Example 5. Explain why $(-3)(5) = -15$.

Solution: This product must be equal to -15 because, when we add 15 to it, we get 0.

$$\begin{aligned} (-3)(5) + 15 &= (-3)(5) + (3)(5) &\quad& [15 = (3)(5)] \\ &= (-3 + 3)(5) &\quad& \text{(3B)} \\ &= (0)(5) &\quad& (-3 \text{ and } 3 \text{ are additive inverses)} \\ &= 0. &\quad& (0 \text{ is an annihilator)} \end{aligned}$$

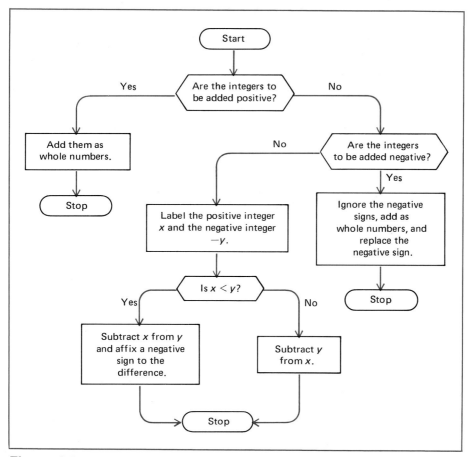

Figure 6.2

Example 6. Explain why $(-3)(-5) = 15$.

Solution: When we add -15 to this product we get 0; therefore, this product must be equal to 15.

$$
\begin{aligned}
(-3)(-5) + -15 &= (-3)(-5) + (-3)(5) \quad &\text{(From Example 5)} \\
&= (-3)(-5 + 5) \quad &\text{(3A)} \\
&= (-3)(0) \quad &\text{(-5 and 5 are} \\
& &\text{additive inverses)} \\
&= 0. \quad &\text{(Zero is an annihilator)}
\end{aligned}
$$

Finally, let us note that the product of two integers is 0 if and only if at least one of the integers is itself equal to 0. That is, $nm = 0$ if and only if either $n = 0$ or $m = 0$.

Now, there are two statements about subtraction in the system of integers; either one of them can be taken as the definition of the operation. The first of these resembles the definition of subtraction of whole numbers:

If x and y are integers, then by x − y we mean that integer d such that x = y + d.

The second statement is

If x and y are integers, then x − y = x + (−y). That is, the result of subtracting y from x is the same as the result of adding the additive inverse of y to x.

If you prefer to take the first statement as the *definition* of subtraction, then the second statement becomes a *theorem* about subtraction. If the second statement is taken as the definition, then the first becomes a theorem. From a mathematical point of view it doesn't matter at all which statement is chosen to be the definition. Exercises 7 and 8 outline proofs of each of these statements upon the assumption of the other as the definition.

A remark is appropriate here on the two meanings of the symbol "−" as it is used in the statement "$x − y = x + (−y)$." On the left-hand side of this equation, the symbol "−" is being used to indicate the subtraction operation. In this usage, the symbol is usually rendered into English as "minus." On the right-hand side, the same symbol is being used as a part of the symbol "$−y$," which denotes the additive inverse of the number y. Used in this way, the symbol is usually rendered into English as either "negative" or the "inverse of."

Example 7. $(−3) − (7) = ?$ $(−3) − (−7) = ?$

Solution: According to the second of the preceding two statements, $(−3) − (7) = (−3) + (−7)$, which is equal to $−10$. Similarly, $(−3) − (−7) = (−3) + (−(−7))$, which is the same as $(−3) + 7$ or 4.

Because every integer has an additive inverse, the symbols x and $(−y)$ have meaning for all integers x and y. Therefore, because addition is a binary operation, the symbol $x + (−y)$ has meaning for all integers x and y. Then, because $x + (−y) = x − y$, the symbol $x − y$ has meaning for all integers x and y. Thus we see that because every integer has an additive inverse, subtraction in the system of integers is a binary operation. This is the single most significant property of the system of integers and is the reason for the usefulness of this number system.

Finally, subtraction shares both the right and left distributive properties with multiplication:

$$x \cdot (y − z) = (x \cdot y) − (x \cdot z) \quad \text{and} \quad (x − y) \cdot z = (x \cdot z) − (y \cdot z).$$

Exercises 6.1

1. What are the additive inverses of 18, -123, 67, 0, and -111?
2. As in Example 1 show why the equation $-5 + -2 = -7$ is true.
3. As in Example 2 show why the equation $11 + -4 = 7$ is true.
4. As in Example 3 show why the equation $-111 + 63 = -48$ is true.
5. As in Example 4 show why the product of -6 and 0 is 0.
6. As in Example 5 show why $(-4)(5) = -20$. Then, as in Example 6, show why $(-4)(-5) = 20$.
7. Define $x - y$ to be that integer d such that $x = y + d$. Justify each step in the following proof that $x - y = x + (-y)$. We begin by putting $x - y = d$. Our task is to prove that $d = x + (-y)$.

Statements	Justifications
1. $x - y = d.$	1. We are using d as a name for the difference.
2. $x = y + d.$	2. By the definition.
3. $x + (-y) = (y + d) + (-y).$	3.
4. $x + (-y) = (d + y) + (-y).$	4.
5. $x + (-y) = d + [y + (-y)].$	5.
6. $x + (-y) = d + 0.$	6.
7. $x + (-y) = d.$	7.
8. $x - y = x + (-y).$	8.

8. Define $x - y$ to be $x + (-y)$. Justify the steps in the following proof that if $x - y = d$, then $x = y + d$.

Statements	Justifications
1. $x - y = d.$	1. Hypothesis.
2. $x + (-y) = d.$	2. By the definition.
3. $[x + (-y)] + y = d + y.$	3.
4. $x + [(-y) + y] = d + y.$	4.
5. $x + 0 = d + y.$	5.
6. $x = d + y.$	6.
7. $x = y + d.$	7.

9. Solve these equations.
   (a) $x = 6 - (-4).$
   (b) $2x = (-4) - (-6).$
   (c) $x = 3 - [-4 - (-5)].$
10. Use the integers -4, 5, and -3 to demonstrate that addition and multiplication share both the right and left distributive properties.
11. The relation *is less than* is defined for integers just as it was for whole numbers: $x < y$ means that there exists a positive integer p such that $x + p = y$. Use this definition to prove that
    (a) $-3 < -2$ (b) $-1000 < -10.$
12. Let x, y, and z be integers and suppose that $x < y$.
    (a) If $z > 0$, then how are xz and yz related?
    (b) If $z < 0$, then how are xz and yz related?

13. Make a flow chart that shows how to determine which of two integers is the smaller.

14. Make a flow chart for performing multiplication of integers.

*15. If we think of the integers as located on the number line, we can define the **absolute value** of an integer to be the distance of that integer from the origin. The absolute value of an integer x is denoted by $|x|$. For example, $|-8| = 8$, since -8 is located 8 units from the origin.

    (a) What are the absolute values of the integers -66, 45, 0, -1, -23, and 14?

    (b) For what values of x is the equation $|-3| \cdot |x| = 6$ true?

    (c) If a positive integer and a negative integer are to be added, and if the absolute value of the negative integer is larger than the absolute value of the positive integer, then is the sum of the integers positive or negative?

    (d) Make a flow chart that shows how to find the absolute value of an integer.

16. The set of all integers can be described using the bracket notation as $\{0, 1, -1, 2, -2, 3, -3, 4, -4, 5, -5, 6, -6, \ldots\}$. Use this listing of the integers to set up a one-to-one correspondence between the set of integers and the set of whole numbers. What then is the cardinal number of the set of integers? Are there more integers than there are whole numbers?

6.2   Algebraic Structure: Groups

It was an Englishman, George Boole (1815–1864), who first clearly saw that mathematics might concern itself with properties of mathematical systems rather than just with properties of the objects of those systems. His idea, quickly followed by similar insights from other mathematicians, was that "pure" mathematics should concern itself with the study of the form of mathematical systems rather than with the content of such systems. Bertrand Russell has called this new understanding of the purpose of mathematical investigation the single greatest achievement of pure mathematics in the nineteenth century. In this section we shall discuss these ideas.

    Suppose that S is a set of objects and that $*$ is a binary operation defined on the objects of S possessing the following properties:

    1. (A) $*$ is commutative.
        (B) $*$ is associative.
        (C) There is a $*$-identity $i \in S$. That is, there is a special object i belonging to the set S such that $x * i = i * x = x$ for every $x \in S$.
        (D) Each object in S has a $*$-inverse. That is, corresponding to each object $x \in S$, there is an object $\tilde{x} \in S$ such that $x * \tilde{x} = \tilde{x} * x = i$.

Then the set S, together with the binary operation *, is called a **commutative group.** We have already seen many examples of sets equipped with binary operations that are commutative groups.

Example 1.   The set of integers, together with the binary operation of addition, is a commutative group. The identity object is 0, and the inverse of an object x is $-(x)$.

Example 2.   The set $\{0, 1, 2, 3, 4, 5\}$, together with the binary operation called 6-addition (see Section 2.6), is a commutative group. The operation of 6-addition is both commutative and associative, and the 6-additive identity is the object 0. Each object in the set has a 6-additive inverse. The 6-additive inverse of 0 is 0, of 1 is 5, of 2 is 4, of 3 is 3, of 4 is 2, and of 5 is 1.

Example 3.   The set of whole numbers, together with the operation of addition, is not a commutative group. The operation is commutative and associative and there is an identity, but the operation fails to possess property 1D. Indeed, it was the need to obtain a number system with an addition operation that would possess property 1D which required us to introduce the negative integers.

Example 4.   Let the set S be the set $\{a, b, c, d\}$ and let a binary operation * be defined by means of the operation table in Table 6.1. In the table the entry in row x and column y is the object $x * y$. It can be shown that this table defines an operation that is both commutative and associative. The *-identity is the object a, since $x * a = a * x = x$ for $x \in S$. (Examine the table to verify this.) The *-inverse of a is a itself, of b is c, of c is b, and of d is d, since $a * a = a$, $b * c = c * b = a$, and $d * d = a$. Hence this four-object set, together with the operation defined by means of the table, is a commutative group.

Example 5.   Let a point P be fixed in the plane. Then, corresponding to each angle α there is a rotation r_α of the plane about the point P through α degrees. If $\alpha > 0$, then the rotation r_α is in the counterclockwise direction. If $\alpha < 0$, then the rotation is in the clockwise direction. Now we can define a binary operation

Table 6.1

		Columns		
*	a	b	c	d
a	a	b	c	d
b	b	d	a	c
c	c	a	d	b
d	d	c	b	a

Rows

(symbolized by *) on the set S of all such rotations as follows: Given two rotations r_α and r_β, the result of operating upon them (in the given order) is the rotation $r_{\alpha+\beta}$. This operation can be described by means of the equation

$$r_\alpha * r_\beta = r_{\alpha+\beta}$$

and is called **composition of rotations**. The set S of all rotations, together with this binary operation *, is a commutative group.

1. (A) * is commutative. For if r_α and r_β are any two rotations, then
$r_\alpha * r_\beta = r_{\alpha+\beta} = r_{\beta+\alpha} = r_\beta * r_\alpha$.
   (B) * is associative. For if r_α, r_β, and r_γ are any three rotations, then $r_\alpha *$
$(r_\beta * r_\gamma) = r_\alpha * r_{\beta+\gamma} = r_{\alpha+(\beta+\gamma)} = r_{(\alpha+\beta)+\gamma} = r_{\alpha+\beta} * r_\gamma = (r_\alpha * r_\beta) * r_\gamma$.
   (C) The *-identity is the rotation through $0°$, r_0:$r_0 * r_\alpha = r_\alpha * r_0 = r_\alpha$.
   (D) The *-inverse of a rotation r_α is the rotation $r_{-\alpha}$, since $r_\alpha * r_{-\alpha} =$
$r_{\alpha+-\alpha} = r_0$ and $r_{-\alpha} * r_\alpha = r_{-\alpha+\alpha} = r_0$.

We worked with the objects of this commutative group in our discussion of the congruence and similarity relations in Chapter 5.

By a **group** we mean a set S, together with a binary operation * that possesses properties 1B, 1C, and 1D. We have not seen any examples of groups that are not also commutative groups, but such noncommutative groups abound in mathematics. Our concern, however, will be only with commutative groups.
   The notion of a group was studied by a young Frenchman, Evariste Galois (1811–1832). Although the idea that mathematics could deal with such abstract ideas was not original with him, he was the first to discuss the notion specifically. (He died in a duel at the age of 21.) The properties that define a group are called structural properties. Very roughly speaking, by a **structural property** of a mathematical system, we mean a property of that system which has nothing whatsoever to do with the particular nature of the objects of that system. For example, the property of being a commutative group is a structural property of a mathematical system. The study of such properties is called **modern** or **abstract algebra**. The abstract algebraist is concerned only with structural properties of mathematical systems and is not at all concerned with the specific nature of the objects of the systems that he studies. If two systems possess exactly the same structural properties, they are called **isomorphic systems.**

Exercises 6.2

1. Consider the set $\{\alpha, \beta\}$, together with the binary operation defined by the following table:

*	α	β
α	α	β
β	β	α

(a) Is the operation commutative? What is the *-identity? What is the *-inverse of α? What is the *-inverse of β?

(b) In order to establish that this operation is associative, it is necessary to verify that each of the following equations is true. Do so by direct computation.

$$(\alpha * \alpha) * \beta = \alpha * (\alpha * \beta) \qquad (\alpha * \beta) * \beta = \alpha * (\beta * \beta)$$
$$(\alpha * \beta) * \alpha = \alpha * (\beta * \alpha) \qquad (\beta * \alpha) * \beta = \beta * (\alpha * \beta)$$
$$(\beta * \beta) * \alpha = \beta * (\beta * \alpha) \qquad (\beta * \alpha) * \alpha = \beta * (\alpha * \alpha)$$
$$(\alpha * \alpha) * \alpha = \alpha * (\alpha * \alpha) \qquad (\beta * \beta) * \beta = \beta * (\beta * \beta).$$

Conclude that this set, together with the binary operation * is a commutative group.

2. In Exercise 1 we saw an example of a group that contained only two objects. It was not an accident that this group was commutative, for it can be proved that any group that contains just two objects must be a commutative group. Can you prove this?

3. In Example 1 we saw a table defining a binary operation that makes the set $\{\alpha, \beta\}$ into a group. Do any of the following tables define binary operations that make $\{\alpha, \beta\}$ into a group? (*Hint:* You may want to use the result of Exercise 2.)

◆	α	β
α	α	α
β	α	β

▲	α	β
α	α	α
β	β	α

●	α	β
α	α	α
β	β	β

■	α	β
α	α	β
β	β	β

4. The set $\{a, b, c, d, e, f\}$, together with the operation defined by means of the table on p. 170, is a group.
   (a) Use the two triples d, b, and c and a, f, and b to illustrate the associativity of this operation.
   (b) Find the *-identity and find the *-inverse of each object of the group.
   (c) Prove by means of a counterexample that this group is not commutative.

5. It is a theorem that if a set S of objects, together with a binary operation *, is a group, then the operation * has the right cancellation property: $x * z = y * z$ implies $x = y$. Prove this theorem. (*Hint:* Consider somehow using the *-inverse of z.)

*	a	b	c	d	e	f
a	a	b	c	d	e	f
b	b	c	a	f	d	e
c	c	a	b	e	f	d
d	d	e	f	a	b	c
e	e	f	d	c	a	b
f	f	d	e	b	c	a

6. Suppose we were to change the entry in row c and column d of the operation table in Exercise 4 by replacing e by c. By using Exercise 5, explain why the operation defined by the changed table would not make the set $\{a, b, c, d, e, f\}$ into a group. (*Hint:* If it were still a group, then you could obtain a contradiction to Exercise 5.)

6.3 The System of Rational Numbers

The notion of a positive rational number is a very ancient one. As early as about 2000 B.C. the Egyptians were working with positive rational numbers whose numerators were equal to 1, and the Chinese were involved with such numbers as $247\frac{933}{1460}$. It would appear that such positive rational numbers were regarded as being much more natural than the negative integers. At any rate, they were accepted as legitimate numbers long before negative integers were. In this chapter we shall augment the system of integers with such numbers, and the number system we obtain as a result is called "the system of rational numbers."

In Section 6.1 we defined the integers to be those numbers that are solutions of equations of the form $x + a = 0$, where a is a nonzero whole number. In a similar way we now define the **rational numbers** to be those numbers that are solutions of equations of the special form $b \cdot x = a$, where a and b are integers and b is different from zero. The solution of this equation is symbolized by a/b. For example, the rational number that is the solution of the equation $2x = 1$ is symbolized by the familiar symbol $\frac{1}{2}$ and, being the solution of the equation $2x = 1$, is that number which, when multiplied by 2, yields 1. The rational number whose symbolic name is $\frac{2}{3}$ is the number which, when multiplied by 3, yields 2; that is, $\frac{2}{3}$ names the solution of the equation $3x = 2$. The solution of the equation $5x = -2$ is symbolized by the symbol $\frac{-2}{5}$ and is that number which, when multiplied by 5, yields -2.

We can also think of rational numbers as being numbers associated with certain points of the number line. For example, suppose that the unit distance (the distance between the origin and the unit) is subdivided into three equal parts and that two of these equal parts are measured off to the right of the origin. The point so determined is associated with the rational number $\frac{2}{3}$ (refer to Figure 6.3). Thus $\frac{2}{3}$ can be regarded (in terms of the number line) as that point which lies to the right of the origin a distance equal to two of the three equal parts of the unit distance. If these two equal parts are measured off to the left of the origin, the point so determined is the point associated with the rational number $\frac{-2}{3}$.

It should be clear that every integer is also a rational number. For example, the integer -6 is a rational number, since -6 is the solution of the equation $1x = -6$ and, by definition, the solution of this kind of equation is a rational number. Thus the set of integers is a proper subset of the set of rational numbers.

The most commonly used symbols for representing the whole numbers are the modern versions $(0, 1, 2, 3, 4, 5, \ldots)$ of the original Hindu–Arabic numerals. The most commonly used symbols for the integers are derived from the whole number symbols: $\ldots, -3, -2, -1, 0, 1, 2, 3, \ldots$. These two methods of symbolizing these numbers are advantageous because there is a *one-to-one correspondence* between the numbers and the symbols that name them. This means that each symbol names just one number and each number is named by just one symbol.

However, the situation is quite different with the rational numbers. The most commonly used names for rational numbers are symbols of the form

$$a/b \quad \text{or} \quad \frac{a}{b},$$

where a and b are integer symbols and b is different from zero. (Recall that the integer a is called the numerator and b is called the denominator.) It is perhaps only a historical accident that we use these two-part symbols, called **fractions,** for representing the rational numbers, but we would be hard pressed to find a better way to name these numbers. In a way, this method of symbolizing rational numbers seems to have been inevitable. But it is not a completely satisfying system, and the reason for this is that there is not a one-to-one correspondence between the numbers and the symbols that name them. For example, the rational number that is named by $\frac{1}{2}$ is also named by each of the infinitely many different fractions

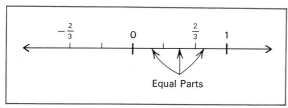

Equal Parts

Figure 6.3

$\frac{-1}{-2}$, $\frac{2}{4}$, $\frac{-2}{-4}$, $\frac{3}{6}$, $\frac{-3}{-6}$, $\frac{4}{8}$, $\frac{-4}{-8}$, and so on. In general, two fractions a/b and c/d name the same rational number if and only if $ad = bc$. We write the sentence $a/b = c/d$ to express the fact that both of these fractions name the same rational number.

Suppose that a/b names a certain rational number and c is a nonzero integer. Then a/b and ac/bc name the same rational number ($a/b = ac/bc$) because $(a)(bc) = (b)(ac)$. This means that we can multiply or divide both the numerator and the denominator of a fraction by the same nonzero integer without changing the meaning of the fraction. This fact has three important applications to the rewriting of fractions.

1. *To rewrite fractions with positive denominators.* For example, $\frac{5}{-4}$ can be written with a positive denominator by multiplying both numerator and denominator by the integer -1: $\frac{5}{-4} = \frac{-5}{4}$.

2. *To rewrite fractions in lowest terms.* A fraction is said to be written in lowest terms when its numerator and denominator are relatively prime. That is, a fraction is in lowest terms when the greatest common divisor of its numerator and denominator is 1 (for example, $\frac{2}{3}$ and $\frac{4}{5}$ are in lowest terms but $\frac{4}{14}$ and $\frac{-6}{9}$ are not). To write a fraction in lowest terms, divide both numerator and denominator by the GCD of these integers.

3. *To rewrite two or more fractions so that they have the same denominator.* The easiest way to do this is to compute the least common multiple of all the denominators and then to multiply both numerator and denominator of each fraction by an appropriate integer so that the denominator will equal this least common multiple. For example, to rewrite the fractions $\frac{7}{18}$ and $\frac{5}{24}$ so that they have the same denominator, compute the LCM of the denominators $2 \cdot 3^2$ and $2^3 \cdot 3$. This LCM is $2^3 \cdot 3^2$. Then

$$\frac{7}{2 \cdot 3^2} = \frac{7 \cdot 2^2}{(2 \cdot 3^2)(2^2)} = \frac{7 \cdot 2^2}{2^3 \cdot 3^2} = \frac{28}{72}$$

and

$$\frac{5}{2^3 \cdot 3} = \frac{5 \cdot 3}{(2^3 \cdot 3)(3)} = \frac{5 \cdot 3}{2^3 \cdot 3^2} = \frac{15}{72}.$$

Up to now we have been careful to distinguish between rational numbers and the fractions that name them, but we shall fail to make this distinction in what follows unless there is a good reason for making it. Thus we may speak of "the rational number $\frac{2}{3}$" or "the numerator of the rational number" when we really mean "the rational number named by the fraction $\frac{2}{3}$" and "the numerator of the fraction that names the rational number."

Before going on to discuss the important operations in this number system, let us define the inequality relations. If two rational numbers are to be compared using the inequality relations, first rename these numbers so that they have the same positive denominator. Suppose that a/b and c/b are two such numbers. Then

$a/b < c/b$ if and only if $a < c$. Also $a/b > c/b$ if and only if $a > c$. The reason the common denominator must be positive is that this definition gives absurd results otherwise. For example, if we disregarded the requirement of a positive denominator, then $-4 < -3$ would imply that $-4/-1 < -3/-1$, from which it would follow that $4 < 3$, which is absurd.

The binary operations of addition and multiplication that are defined in the system of rational numbers are extensions of the corresponding operations in the system of integers in the sense that, given two integers, we can add or multiply them either by using addition or multiplication in the system of integers or by using addition or multiplication in the system of rational numbers. We get the same sum or product no matter which we do. Addition of rational numbers is easy to perform once the summands have been renamed so that their denominators are the same. We use the rule

$$\frac{a}{b} + \frac{c}{b} = \frac{a + c}{b}$$

to perform addition. (See Figure 6.4.) Multiplication is even easier, because no renaming is necessary before the operation can be applied to a pair of rational numbers:

$$\frac{a}{b} \cdot \frac{c}{d} = \frac{ac}{bd}.$$

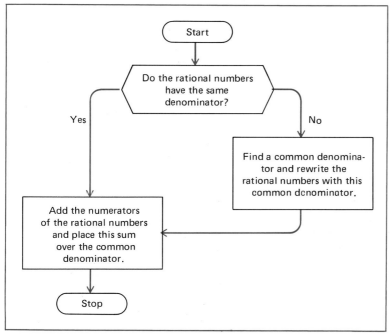

Figure 6.4

It can be proved that these two operations possess the same properties as addition and multiplication of integers. Thus addition has (see p. 50) properties 1A (commutivity), 1B (associativity), 1C (identity), and 1D (inverses). Multiplication has properties 2A (commutivity), 2B (associativity), and 2C (identity). The two operations together possess the two distributive properties 3A and 3B. In addition to these, multiplication possesses one additional property of great importance that distinguishes the rational number system from the other number systems we have studied.

    2D. Every nonzero rational number possesses exactly one inverse with respect to multiplication.

By this we mean that, given any nonzero rational number x, there is a rational number y, called the **multiplicative inverse** of x, such that $x \cdot y = y \cdot x = 1$. For example, the multiplicative inverses of $\frac{3}{4}$, $\frac{-5}{4}$, and 9 are $\frac{4}{3}$, $\frac{-4}{5}$, and $\frac{1}{9}$, respectively. We shall denote the multiplicative inverse of a nonzero rational number x by the symbol $(x)^{-1}$. Thus $(\frac{3}{4})^{-1} = \frac{4}{3}$, $(\frac{-5}{4})^{-1} = \frac{-4}{5}$, and $(9)^{-1} = \frac{1}{9}$. Zero has no multiplicative inverse.

    Subtraction of rational numbers is performed upon rational numbers that have already been renamed so that they have the same common denominator according to the rule

$$\frac{a}{b} - \frac{c}{b} = \frac{a-c}{b}.$$

If a/b and c/b are any two rational numbers that have been written with the same denominator, then since the symbol $a - c$ represents a unique integer, the symbol $(a - c)/b$ names a unique rational number. This means that subtraction in the system of rational numbers is a binary operation.

    We are now ready to define division of rational numbers. There are two ways that we could define division. Whichever of these two statements is used to define division, the other takes on the status of a theorem in the logical development of the theory. Here are the statements.

1. *If a/b and c/d are fractions, and if c/d is not equal to 0, then the symbol $a/b \div c/d$ names that rational number q such that $a/b = (c/d) \cdot q$. That is, $a/b \div c/d = q$ if and only if $a/b = (c/d) \cdot q$.*
2. *If a/b and c/d are fractions, and if c/d is not 0, then $a/b \div c/d = (a/b) \cdot (c/d)^{-1}$.*

If we adopt statement 1 as the definition of division of rational numbers, here is a proof of statement 2.

Theorem.

$$\frac{a}{b} \div \frac{c}{d} = \left(\frac{a}{b}\right) \cdot \left(\frac{c}{d}\right)^{-1}.$$

Proof:

Suppose that a/b and c/d are rational numbers and that c/d is different from 0. Let

$$\frac{a}{b} \div \frac{c}{d} = q.$$

Then, according to our definition of division,

$$\frac{a}{b} = \frac{c}{d} \cdot q.$$

Since $c/d \neq 0$, $(c/d)^{-1}$ exists and we can write

$$\frac{a}{b} \cdot \left(\frac{c}{d}\right)^{-1} = \left(\frac{c}{d} \cdot q\right) \cdot \left(\frac{c}{d}\right)^{-1}$$

$$= \left(q \cdot \frac{c}{d}\right) \cdot \left(\frac{c}{d}\right)^{-1}$$

$$= q \cdot \left(\frac{c}{d} \cdot \left(\frac{c}{d}\right)^{-1}\right)$$

$$= q \cdot 1$$

$$= q.$$

Combining the first and last equations, we get the statement of the theorem and the proof is complete.

Division by nonzero rational numbers is always possible; this is the most significant property of the system of rational numbers. Suppose that a/b and c/d are two rational numbers and $(c/d) \neq 0$. Then the multiplicative inverse of c/d exists and, since multiplication is a binary operation, $(a/b) \cdot (c/d)^{-1}$ has a unique meaning. According to the last theorem, $(a/b) \cdot (c/d)^{-1}$ has exactly the same meaning as $(a/b) \div (c/d)$. Hence $(a/b) \div (c/d)$ has a unique meaning and division by nonzero rational numbers is always possible; the result is always a unique rational number. Although division in this number system is not a binary operation (since division by zero is not defined), it comes as near to being one as any division process can come.

Exercises 6.3

1. Draw a number line and locate the rational numbers $\frac{3}{4}$, $\frac{5}{4}$, $\frac{1}{7}$, $\frac{-3}{7}$, and $\frac{16}{7}$. Then locate the additive inverses of these numbers. Finally, locate the multiplicative inverses.

2. Write the rational numbers $\frac{4}{6}, \frac{-16}{24}, \frac{198}{144}$, and $(2^2 \cdot 3^4 \cdot 5 \cdot 7^3 \cdot 11)/(2^3 \cdot 3^2 \cdot 5^2 \cdot 7 \cdot 13)$ in lowest terms.

3. Solve these equations.
   (a) $2x - \frac{3}{4} = \frac{5}{7}$.　　　　(b) $x = \frac{2}{3} \div (\frac{1}{2} - \frac{-2}{3})$.
   (c) $\frac{4}{7}x - \frac{1}{3} = \frac{1}{2}$.　　　　(d) $2x = (\frac{3}{4} \div \frac{-3}{4}) \div \frac{1}{5}$.

4. Make flow charts for subtraction, multiplication, and division of rational numbers.

5. Arrange these numbers from left to right in order of increasing size; $\frac{2}{3}, \frac{-2}{3}$, $\frac{4}{5}, \frac{4}{-5}, \frac{1}{5}, \frac{-1}{9}, 0, \frac{1}{2}, \frac{-1}{16}, -7$, and 12.

6. Is there a smallest positive rational number? Is there a smallest positive integer? Is there a largest negative rational number? Is there a largest negative integer?

7. Prove that between every pair of distinct rational numbers there is a third rational number. (*Hint:* Consider two rational numbers a/b and c/d with $a/b < c/d$. Average these numbers and show that this average is greater than a/b and less than c/d.)

8. Complete the following.
   (a) $a/b + c/d = ?/bd$.　　　　(b) $a/b - c/d = ?/bd$.

*9. The Egyptians worked only with positive rational numbers whose numerators were equal to 1. Use the fact that

$$\frac{1}{n} = \frac{1}{n+1} + \frac{1}{n(n+1)}$$

to write the following rational numbers as sums of distinct rational numbers whose numerators are equal to 1. [*Hint:* $\frac{2}{3} = \frac{1}{3} + \frac{1}{3} = \frac{1}{3} + (\frac{1}{4} + \frac{1}{12}) = \frac{1}{3} + \frac{1}{4} + \frac{1}{12}$.]
   (a) $\frac{2}{7}$.　　　(b) $\frac{3}{7}$.　　　(c) $\frac{4}{5}$.

10. Your answers should make it clear why the Egyptians were able to work only very simple problems involving rational numbers. As simple a number as $\frac{4}{5}$ must have been regarded by them as a most complicated number because their name for it was the sum of 15 numbers with numerators equal to 1.
   Prove the fact used in the preceding exercise:

$$\frac{1}{n} = \frac{1}{n+1} + \frac{1}{n(n+1)}$$

by adding the summands on the right side and simplifying.

*11. We have seen that the cardinal number of the set of integers is $\aleph_0$ so that there are just as many whole numbers as there are integers. Even more surprising is the fact that the set of positive rational numbers also has cardinal number $\aleph_0$. In other words, there are no more positive rational numbers than there are whole numbers. (Incidentally, it can be proved from this that the cardinal number of the set of all rational numbers is also $\aleph_0$.) Here is the outline of the proof of this fact.

(a) Arrange the fractions in an "infinite" array as shown in the figure. Explain why every fraction representing a positive rational number occurs somewhere in this array of fractions. Where would you find the fraction $\frac{56}{78}$ in this array?

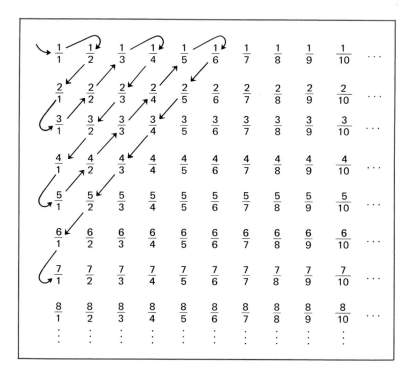

(b) By tracing along the path as indicated, we set up a correspondence between the whole numbers and the positive rationals that looks like this for small whole numbers:

$$
\begin{array}{ccccccccccccc}
0 & 1 & 2 & 3 & 4 & 5 & 6 & 7 & 8 & 9 & 10 & 11 & 12 & \cdots \\
\updownarrow & \updownarrow & \updownarrow & \updownarrow & \updownarrow & \updownarrow & \updownarrow & \updownarrow & \updownarrow & \updownarrow & \updownarrow & \updownarrow & \updownarrow \\
1 & \frac{1}{2} & 2 & 3 & \frac{1}{3} & \frac{1}{4} & \frac{2}{3} & \frac{3}{2} & 4 & 5 & \frac{1}{5} & \frac{1}{6} & \frac{2}{5} & \cdots
\end{array}
$$

As we moved along the path, we occasionally skipped a positive rational. Why? Explain why our procedure sets up a pairing between the whole numbers and the positive rational numbers that is a one-to-one correspondence. Find the positive rational numbers that correspond to the whole numbers 13 through 50.

6.4 Algebraic Structure: Fields

In this section we shall continue our discussion of algebraic structure of mathematical systems by defining the concept of a field.

The set of rational numbers together with the binary operation of addition is a commutative group, but this number system has more structure than this. The subset of objects different from the additive identity 0 forms a commutative group with respect to multiplication. Also, the addition and multiplication operations are connected by means of the distributive properties. Any mathematical system with these properties is called a field. A **field,** then, is a mathematical system consisting of a set S of objects, together with two binary operations $*$ and $\circ$, which possess the following properties:

1. (A) $*$ is commutative.
   (B) $*$ is associative.
   (C) There is a $*$-identity i in the set S: $x * i = i * x = x$.
   (D) Each object $x \in S$ has a $*$-inverse $\tilde{x}$ in the set S: $x * \tilde{x} = \tilde{x} * x = i$.
2. (A) $\circ$ is commutative.
   (B) $\circ$ is associative.
   (C) There is a $\circ$-identity e (different from i) in the set S: $x \circ e = e \circ x = x$.
   (D) Each object $x \in S$ different from i has a $\circ$-inverse $\bar{\bar{x}}$ in the set S:
   $x \circ \bar{\bar{x}} = \bar{\bar{x}} \circ x = e$.
3. These left and right distributive properties connecting the operations $*$ and $\circ$ hold:
   (A) $x \circ (y * z) = (x \circ y) * (x \circ z)$.
   (B) $(x * y) \circ z = (x \circ z) * (y \circ z)$.

As we have just said, the system of rational numbers possesses each of these properties (when S means the set of rational numbers $*$ means addition and $\circ$ means multiplication); thus, the system of rational numbers is a field. Here is an example of a field containing only a finite number of objects.

Example 1.   The system of arithmetic modulo five is a field. The objects of this system are the objects of the set $\{0, 1, 2, 3, 4\}$ and the operations are 5-addition ($\oplus$) and 5-multiplication ($\otimes$). These operations are performed upon two objects by first adding or multiplying the objects, using ordinary addition or multiplication of whole numbers and then casting out fives from the resulting sums or products. What remains after fives have been cast out is the 5-sum or 5-product of the objects. For example, $3 \oplus 4 = 2$ and $3 \otimes 4 = 2$. The operations $\oplus$ and $\otimes$ are both commutative and associative, and they share the distributive properties. (We could verify these statements, but we shall not bother to do so.) The $\oplus$-identity is 0 and the $\otimes$-identity is 1. The $\oplus$-inverses of all objects and the $\otimes$-inverses of objects different from the $\oplus$-identity are listed in Table 6.2.

Since all the field properties hold in this system, arithmetic modulo five is a field.

Table 6.2

Object	⊕-Inverse	⊗-Inverse
0	0	—
1	4	1
2	3	3
3	2	2
4	1	4

Example 2. The system called arithmetic modulo six is not a field because not every object (different from the ⊕-identity 0) has a ⊗-inverse. For example, 3 has no ⊗-inverse in this system ($3 \otimes 0 = 0$, $3 \otimes 1 = 3$, $3 \otimes 2 = 0$, $3 \otimes 3 = 3$, $3 \otimes 4 = 0$, $3 \otimes 5 = 3$ and so there is no object $\bar{x}$ in the system such that $3 \otimes \bar{x} = \bar{x} \otimes 3 = 1$, and therefore 3 has no ⊗-inverse). What is the difference between arithmetic modulo five and arithmetic modulo six? Why should one of them be a field and the other not? The answer lies in the fact that 5 is a prime but 6 is composite. It can be proved that if p is a prime number, then arithmetic modulo p is a field; but if c is composite, then arithmetic modulo c is not a field.

Example 3. The system of integers is not a field because property 2D fails to hold in this system. It was to obtain a number system possessing property 2D that the rational numbers were introduced.

There is another structural property possessed by the field of rational numbers called "the order property." This property can be phrased for an arbitrary field, but let us be content with its statement for the field of rational numbers.

4. The field of rational numbers is equipped with an order relation $<$ such that
   (A) $0 < x$ and $0 < y$ imply $0 < x + y$ and $0 < x \cdot y$.
   (B) If $0 < x$, then $-(x) < 0$.
   (C) Exactly one of these statements is true: $0 < x$, $x < 0$, $x = 0$.

Because the system of rational numbers is a field with property 1, we call the field of rational numbers an **ordered field**.

We shall continue our discussion of structural properties of mathematical systems in Section 6.6.

Exercises 6.4 _____

1. Consider the set $\{a, b, c\}$, together with the two operations defined by means of these tables:

*	a	b	c
a	a	b	c
b	b	c	a
c	c	a	b

○	a	b	c
a	a	a	a
b	a	b	c
c	a	c	b

You may assume that each of these operations is associative and that the operations share the two distributive properties given in Property 3. Verify that the rest of the properties (1A, 1C, 1D, 2A, 2C, and 2D) hold and thus that this set, together with these operations, is a field.

2. Arithmetic modulo four (that is, the set $\{0, 1, 2, 3\}$, together with 4-addition and 4-multiplication) is a group (with respect to 4-addition) but is not a field because one of the properties 2A, 2B, 2C, 2D, 3A, 3B fails to hold. Find the property that does not hold for arithmetic modulo four.

3. Does the group of integers possess the order property? The system of whole numbers does not possess this property. Why not?

4. Is it correct to say that property 4C means that either $0 < x$ or $x < 0$ or $x = 0$?

6.5   Decimal Numerals and Rational Numbers

We have been naming rational numbers with fractions. Now we want to discuss the naming of such numbers by decimal numerals.

Given a fraction name of a rational number, it is easy enough to find the decimal numeral name for that number. The procedure is simply to long divide the numerator of the fraction by its denominator. For example, 3 long divided by 8 is .375 and so $\frac{3}{8} = .375$. Such numerals as .375 are called **terminating numerals** because they have only a finite number of digits to the right of the decimal point. But not all rational numbers can be named with terminating numerals. For example, 5 long divided by 6 yields the nonterminating numeral $.83333\cdots$. This numeral is called **nonterminating** because a digit appears in every place to the right of the decimal point. Another example is $\frac{41}{333} = .123123123123\cdots$.

The nonterminating numerals $.83333\cdots$ and $.123123123123\cdots$ are called **repeating numerals** because each consists of a block of digits that is repeated over and over to form the entire numeral. This repeating block does not have to begin repeating at the tenths place (as does $.123123123\cdots$); it may begin to repeat at any place. The numeral $.83333\cdots$ has a repeating block consisting of the single digit 3, which begins to repeat in the hundredths place.

We could examine a great many rational numbers and we would see that in every case the decimal numeral name is either a terminating numeral or a nonterminating but repeating numeral. In fact, the converse of this is true as well. That is, a decimal numeral names a rational number if and only if the numeral is either terminating or nonterminating but repeating.

If we are given the decimal numeral name of a rational number, then we can find a fraction name for that number. This is particularly easy to do if the given decimal numeral is terminating. For example, consider the terminating numeral 1.56. We use the fact that such terminating numerals are really nothing more than abbreviations for the sums of their place values. Thus 1.56 is a symbol that abbreviates the sum $1 + \frac{5}{10} + \frac{6}{100}$ and so

$$1.56 = 1 + \frac{5}{10} + \frac{6}{100} = \frac{100}{100} + \frac{50}{100} + \frac{6}{100}$$

$$= \frac{100 + 50 + 6}{100} = \frac{156}{100} = \frac{39}{25}.$$

In the case of nonterminating numerals, however, the situation is complicated by the fact that, because we cannot add infinitely many numbers, we cannot regard nonterminating numerals as the sum of their place values. We need another procedure in these instances. This procedure is demonstrated in the following two examples.

Example 1. The decimal numeral $.4545454545\cdots$ is nonterminating but repeating, so we know that it names a rational number. Let x represent this rational number so that we can write $x = .4545454545\cdots$. Multiplying both sides of this equation by 100, we get the equation $100x = 45.45454545\cdots$. Now we can subtract these equations,

$$100x = 45.4545454545\cdots$$
$$x = .4545454545\cdots$$
$$\overline{99x = 45.00000\cdots = 45}$$

and see that $99x = 45$ so that $x = \frac{45}{99} = \frac{5}{11}$. Hence $.4545454545\cdots = \frac{5}{11}$.

The length of the repeating block dictates the power of 10 that must be used to multiply the first equation. In Example 1 the length of the block was 2, and so we multiplied the first equation by 100. If the repeating block had had length 3, then we would have multiplied by 1000. Example 1 was fairly simple because the repeating block began to repeat in the tenths place. If such is not the case, then proceed as in the next example.

Example 2. Find a fraction naming the rational number $5.6454545\cdots$.

Solution: Write $x = 5.645454545\cdots$. This numeral does not begin to repeat in the tenths place, so we multiply by 10 and look at $10x = 56.454545\cdots$. Now we have a numeral that begins to repeat in the tenths place. The length of the repeating block is 2, so we multiply this equation by 100 and subtract:

$$
\begin{aligned}
100(10x) &= 5645.45454545\cdots \\
10x &= 56.45454545\cdots \\
\hline
990x &= 5589.0000\cdots = 5589.
\end{aligned}
$$

Hence $x = \frac{5589}{990} = \frac{621}{110}$. That is, $5.645454545\cdots = \frac{621}{110}$.

Exercises 6.5

1. Find fractions that name the rational numbers named by these decimal numerals.
   (a) $.22222\cdots$.   (b) $4.3232323232\cdots$.
   (c) $12.1622222\cdots$.   (d) $.1365757575757\cdots$.

2. A rational number that is represented by a fraction a/b written in lowest terms has a terminating decimal numeral if and only if the prime factorization of b has the special form $2^x \cdot 5^y$. Which of the following have terminating decimal numerals? Verify your answer by finding the decimal numerals of all of these rational numbers.
   (a) $\frac{4}{10}$.   (b) $\frac{4}{9}$.   (c) $\frac{3}{17}$.   (d) $\frac{1}{125}$.

3. Let T denote the set of all rational numbers that can be named by terminating decimal numerals. If $x \in T$ and $y \in T$, is $x + y \in T$? Is $x \cdot y \in T$? Is $x - y \in T$? Is $x \div y \in T$? (*Hint:* The theorem mentioned in Exercise 2 bears on this exercise.)

4. Show that the repeating blocks of the decimal numerals for $\frac{1}{7}$, $\frac{5}{7}$, and $\frac{13}{7}$ all have the same length. This comes as a result of the theorem: If $1/q$ has a repeating block of length n, and if p and q are relatively prime, then p/q also has repeating block of length n.

5. Verify that $\frac{1}{99}$ and $\frac{33}{99}$ have repeating blocks of different lengths. Why is this not a counterexample to the theorem stated in Exercise 4? Can you find another pair of rational numbers whose denominators are the same but that have repeating blocks of different lengths?

6.6   The System of Real Numbers

For the very ancient Greeks (prior to about 700 B.C.), the rational numbers were the only numbers there were, but sometime before about 600 B.C. (these dates are impossible to establish with any kind of accuracy) the existence of numbers that were not rational was discovered. Euclid knew that there were such irrational numbers, but their development was slow; it was not until late in the nineteenth century that they were finally given the careful mathematical treatment they required. This work was done largely by Cantor (whom we met in Chapter 4 in connection with set theory) and the German mathematician J. W. R. Dedekind (1831–1916). These numbers, together with the rational numbers, comprise the most significant number system in mathematics and are the basis for a most important part of modern mathematics, called "analysis." In this section we shall discuss these numbers and the number system to which they give rise.

In Section 6.3 we saw how the rational numbers may be located along the number line. Let us suppose that each time we locate a rational number on the line we color its point. Then we would see that these points in color are strung all along the number line and that there are no segments of the number line which do not contain points in color. This is not particularly difficult to understand. But what is considerably less obvious is that not every point of the number line is in color; that is, there exist points at which no rational number is located. Here is how to find one of these points of the number line that are not in color.

As shown in Figure 6.5, construct a square of side 1 unit distance and place it along the number line. Then the diagonal $\overline{OP}$ decomposes the square into two right triangles. Centering attention on either one of these triangles, we can use the Pythagorean theorem to conclude that the length of the diagonal $\overline{OP}$, d, is given by the equation $d^2 = 1^2 + 1^2$ or $d^2 = 2$. Thus the length of the diagonal $\overline{OP}$ is the number we call the square root of 2 and denote by the symbol $\sqrt{2}$. There is an elementary proof (known to Euclid) that proves that this number, $\sqrt{2}$, is not a rational number. Thus the point P cannot correspond to a rational number and therefore is one of the points that are not in color.

The conclusion we draw from the existence of points that are not in color is that if the only numbers we have available to us are the rational numbers, then there are line segments whose lengths cannot be measured—which do not have a length at all if a length is something that can be measured.[1] This means that

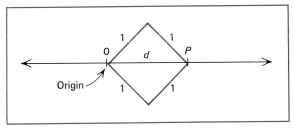

Figure 6.5

---

[1]We could measure the approximate length of the segment $\overline{OP}$ using only rational numbers, but here we are talking about the exact length of this segment.

we need to introduce still more numbers into our number system if all segments are to have measurable length. The number $\sqrt{2}$ is an example of just such a number. We call these new numbers **irrational numbers.** These irrational numbers are the numbers needed to measure the lengths of segments whose lengths cannot be measured using rational numbers. In terms of the number line, they are numbers that correspond to the points that are not in color.

There are three ways to think of an irrational number. We have just identified the irrational numbers as those numbers that correspond to the points on the number line that are not in color—to the points to which no rational number corresponds. We can also think of an irrational number as one that cannot be expressed as the quotient of two integers. A third way to think of an irrational number is by using decimal numerals. We observed in the preceding section that the decimal numeral name for a rational number is either terminating or non-terminating and repeating. It then follows that the decimal numerals that are nonterminating and nonrepeating must name some kind of number other than a rational number. Such numerals name the irrational numbers. Here are some decimal numerals, each of which names an irrational number:

$$1.12345678910111213141516171 8\cdots$$
$$1.01001000100001000001000000 1\cdots$$
$$1.12233344445555566666677777778\cdots.$$

These numerals are nonterminating (obviously) and do not repeat (less obviously) so that each must name an irrational number.

We have stated that the number $\sqrt{2}$ is an irrational number. The fact is that the square root of any positive integer that is not a perfect square is an irrational number. Hence the numbers $\sqrt{2}$, $\sqrt{3}$, $\sqrt{5}$, $\sqrt{6}$, $\sqrt{7}$, $\sqrt{8}$, $\sqrt{10}, \ldots$, are all irrational. The cube root of any positive integer that is not a perfect cube is also irrational, thus $\sqrt[3]{2}$, $\sqrt[3]{3}$, $\sqrt[3]{4}$, $\sqrt[3]{5}$, $\sqrt[3]{6}$, $\sqrt[3]{7}$, $\sqrt[3]{9}, \ldots$ are irrational. In general, the nth root of any positive integer that is not a perfect nth power is irrational.

The number π is irrational. It follows that the decimal numeral for π is nonterminating and nonrepeating. This numeral begins like this:

$$3.14159265358979323846264338327950\cdots.$$

There is no discernible pattern to the occurrence of the digits in this numeral. In recent years computers have been used to find many thousands of the digits in this numeral but the complete numeral can never be known.

By a **real number,** we mean a number that is either rational or irrational. Hence the set of all real numbers is the union of the set of rational numbers and the set of irrational numbers. The set theoretic relationships between the various sets of numbers that we have studied is shown in Figure 6.6.

In our later work, we shall most often want to think of the real numbers as being those numbers that correspond to the points of the number line. It is often

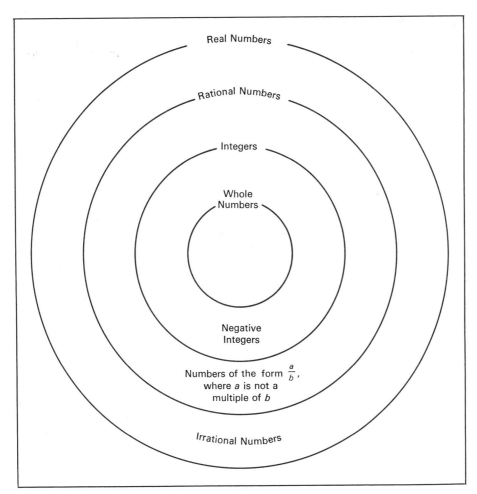

Figure 6.6

convenient to identify the real numbers with their corresponding points and thus to think of the real numbers as actually being points. For this reason we often refer to the number line as the real number line.

Let us remark that there exist numbers that are not real numbers. For example, the number $\aleph_0$ (the smallest infinite cardinal number and the cardinal number of the set of whole numbers) is a number that is not a real number. We can see that $\aleph_0$ is not a real number by observing that there is no point of the number line which is located an infinite distance from the origin, and so no point of the line can be associated with the infinite cardinal number $\aleph_0$. Therefore, $\aleph_0$ is not a real number. There are other kinds of numbers that are not real numbers as well, but we shall not discuss such numbers in this book.

Computations involving real numbers are usually performed by using the decimal numeral names for these numbers, but since all of our computational work in later chapters will involve rational numbers alone, we shall ignore this aspect of the system of real numbers. We should, however, note the structural properties of this system. The set of real numbers, together with the binary operations of addition and multiplication, possess the following properties (see pp. 178 and 179): 1A, 1B, 1C, 1D, 2A, 2B, 2C, 2D, 3A, 3B, 4A, 4B, and 4C. Consequently, the system of real numbers is an ordered field. On the basis of these structural properties alone, we cannot distinguish between the system of rational numbers and the system of real numbers. That is, on the basis of the ordered field properties alone, structurally speaking these two number systems are the same. But there is another structural property possessed by the system of real numbers which is not possessed by the system of rational numbers and which can be used to differentiate between these two number systems. This property is called the **completeness property.** A precise statement of this property in a short space is impossible. The property is definitely a complicated one. It is the basis for differential and integral calculus, as well as most of that portion of modern mathematics called analysis. We must be satisfied with a rough description of this important and powerful property of the real numbers.

Suppose that we locate all the rational numbers on the number line and then throw away all the points of the line that do not correspond to a rational number. For example, we would throw away the point corresponding to $\sqrt{2}$. Then, after we had done this, the line would have gaps or holes in it. In other words, the line would not be *complete*. But if, after locating all the real numbers on the line, we threw away those points not corresponding to any real number, there would be no such gaps or holes in the line because we would not have thrown any points away. That is, the real number line is complete. Every point corresponds to a real number and every real number corresponds to a point—there is a one-to-one correspondence between the set of all points on the line and the set of all real numbers.

We call the field of real numbers a **complete ordered field.** The field of rational numbers is not a complete ordered field. In fact, the only complete ordered field is the field of real numbers. More precisely, it can be proved in advanced courses in abstract algebra that if F is any complete ordered field, then F cannot be distinguished from the field of real numbers on the basis of structural properties alone.

Exercises 6.6

1. The irrational number $\sqrt{5}$ is the length of the diagonal of a rectangle of length 2 and width 1. Use this fact to locate the number $\sqrt{5}$ on the number line. Find the points of the number line that correspond to the irrational numbers $\sqrt{10}$ and $\sqrt{13}$.
2. Describe how you would locate the point $\sqrt{3}$ on the number line, using a right triangle whose hypotenuse has length 2 and one of whose legs has length 1.

3. Prove that there does not exist an isosceles right triangle whose sides are integers.
*4. Here is Euclid's proof of the irrationality of $\sqrt{2}$.

Theorem.
$\sqrt{2}$ is not a rational number (and, therefore, is irrational).

Proof:
(By indirect method.) We begin by assuming that $\sqrt{2}$ is a rational number. Then $\sqrt{2}$ can be named by a fraction. Moreover, like all rational numbers, $\sqrt{2}$ can be named by a fraction that has been written in lowest terms. Let us, therefore, write $\sqrt{2} = a/b$, where a and b are relatively prime. Now explain why these statements follow.

(1) $\sqrt{2} \cdot b = a$.   (2) $2b^2 = a^2$.
(3) a^2 is divisible by 2.   (4) a is divisible by 2.
(5) $a = 2x$.   (6) $2b^2 = 4x^2$.
(7) $b^2 = 2x^2$.   (8) b^2 is divisible by 2.
(9) b is divisible by 2.

Statements 4 and 9 together give us a contradiction. Why? Hence the initial assumption that $\sqrt{2}$ was rational is false; thus $\sqrt{2}$ is not a rational number, and this means that $\sqrt{2}$ is an irrational number.

*5. Repeat the proof given in Exercise 4 making the appropriate changes so as to obtain a proof that $\sqrt{3}$ is irrational.

6. Which of the following are irrational numbers?
    (a) $\sqrt{2} + 1$.   (b) $(\sqrt{2} \cdot \sqrt{2}) + 3$.
    (c) $(\sqrt{2} + 1)(\sqrt{2} - 1)$.   (d) $(\sqrt{2})^2 - 1$.

7. Explain why the decimal numeral $1.010010001000010000010 \cdots$ is nonrepeating. (*Hint:* Assume it is repeating and note that the repeating block, which has some definite length, must contain both 0's and 1's. Argue to a contradiction.)

8. Is the sum of two irrational numbers necessarily irrational? Is the product of two irrational numbers necessarily irrational? (*Hint:* Study Exercise 6.)

9. Complete the table on p. 188 with "yes" or "no."

	Whole Numbers	Integers	Rational Numbers	Real Numbers	Arithmetic Modulo 6	Arithmetic Modulo 5
Addition						
Commutative						
Associative						
Identity						
Inverses						
Multiplication						
Commutative						
Associative						
Identity						
Inverses						
Distributivity of Addition and Multiplication						
$nm = 0$ if and only if $n = 0$ or $m = 0$						
Order Property						
Completeness Property						

Analytic Geometry

In our study of the real numbers we saw that the real numbers may be placed in one-to-one correspondence with the points on a line. This one-to-one correspondence allows us to think of points as being numbers or to think of numbers as being points whenever it is convenient to do so. What we have established is a connection between arithmetic on the one hand and geometry on the other. As tenuous as this connection may appear to be, when fully developed it provides a strong and extremely useful tool for the study of both Euclidean geometry and algebra. The name of this tool is analytic geometry, and we shall be studying some of the basic ideas about this part of mathematics in this chapter.

Analytic geometry stems from the work of the two most important seventeenth century mathematicians, Pierre de Fermat (French, 1601–1665) and René Descartes (French, 1596–1650). Fermat made significant contributions to the subject, but unfortunately he did not publish during his lifetime. Descartes, who probably made his discoveries at roughly the same time as Fermat (about 1629), published a manuscript in which he dealt with connections between algebra and geometry, and it is this work that is usually regarded as the beginning of the subject. As a matter of fact, Fermat's work more nearly resembles modern analytic geometry than does Descartes' work, but, because he published and Fermat did not, it is Descartes' name that is associated with the subject.

7.1 The Cartesian Coordinate System

We are able to locate the solutions of such equations as $3x + 5 = 8$ on the real number line and so give a kind of geometric picture of these solutions. The reason we can do this is that we earlier found a way to attach "number meaning" to each of the points of a line. Before we get involved in analytic geometry, it will be necessary for us to do the same thing in the plane. We must find a way of giving each point of the plane a kind of "number meaning."

The names we shall give to points in the plane are called **ordered pairs** of real numbers. An ordered pair is a two-object set whose two objects are given in a specified order. The ordered pair consisting of the two real numbers 2 and 3, given in that order, is denoted by $(2, 3)$. The symbol $(3, 2)$ means the two-object set consisting of 3 and 2, given in that order. Two ordered pairs (a, b) and (c, d) are the same ordered pair if and only if their left-hand parts are the same and their right-hand parts are the same: $(a, b) = (c, d)$ if and only if $a = c$ and $b = d$.

Here is how we find the ordered name of a point in the plane: Select a pair of perpendicular lines in the plane. For convenience, make one of them vertical and the other horizontal. These lines are called the **coordinate axes.** The horizontal line is called the **x-axis** and the vertical line is called the **y-axis.** Their point of intersection is called the **origin.** Next, using the origin as the point corresponding to the real number 0 and using some convenient unit distance, set up a one-to-one correspondence between the points on each axis and the real numbers. We have done this in Figure 7.1, where we have indicated a few points on the axes that correspond to integers. In effect, we have placed two real number lines in the plane so that they are perpendicular to each other and intersect at their origins. Using these axes, we can assign an ordered pair of real numbers to each point of the plane. If P is a point (refer to Figure 7.2), we construct lines through P perpendicular to the coordinate axes. The left-hand part of the ordered pair name for P is the number assigned to the point on the x-axis determined by the vertical line through P. The right-hand part of the ordered pair name for P is the real number assigned to the point on the y-axis determined by the horizontal line through P. In Figure 7.3 we have located a few points and have given their ordered pair names.

This method for giving ordered pair names to points of the plane sets up a one-to-one correspondence between the set of all points in the plane and the set of all ordered pairs of real numbers. That is, each point has exactly one ordered pair name and each ordered pair names exactly one point. We call the ordered pair name of a point the **coordinate** of that point. The left-hand part of the coordinate is called the **x-coordinate,** and the right-hand part is called the **y-coordinate.**

The coordinates of points in the plane can be used for a great many different purposes, one of which is to find the distance between points. If two points belong to a line that is parallel to one of the coordinate axes, it is very easy to find the distance between them. Consider these examples.

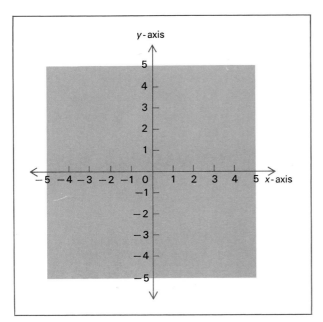

Figure 7.1

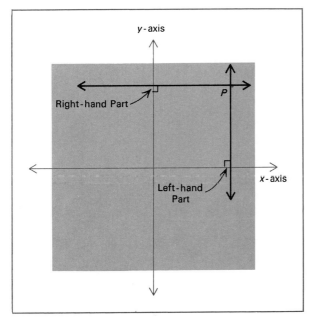

Figure 7.2

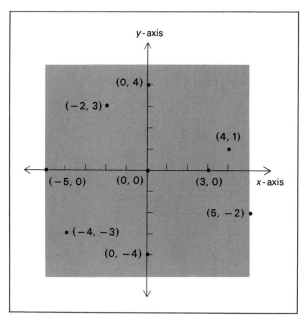

Figure 7.3

Example 1.   Find the distance between the points $(1, 2)$ and $(-2, 2)$.

Solution: These points determine a line parallel to the x-axis (see Figure 7.4). Thus the distance between them is found by subtracting the smaller x-coordinate from the larger x-coordinate. In this case the difference is $1 - (-2)$ or 3. Hence these two points are 3 units apart.

Example 2.   The distance between the points $(3, -3)$ and $(3, 5)$ (see Figure 7.4) is found by subtracting the smaller y-coordinate from the larger y-coordinate. The distance between these points is $5 - (-3) = 8$ units.

   If the line through the two given points is not parallel to a coordinate axis, then the problem of finding the distance between the points is somewhat more complicated. For example, if P and Q are as shown in Figure 7.5, we first construct a right triangle, as shown in that illustration. This right triangle should have the line segment $\overline{PQ}$ as its hypotenuse and should have its legs parallel to the coordinate axes. Since the legs of this triangle are parallel to the coordinate axes, we may find their lengths as in Examples 1 and 2. Once the lengths of these two legs are known, the Pythagorean theorem can be used to find the length of the hypotenuse.

Example 3.   Find the distance between the points $(5, 2)$ and $(1, -1)$.

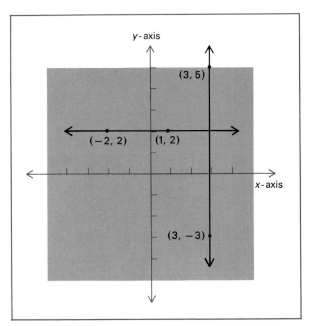

Figure 7.4

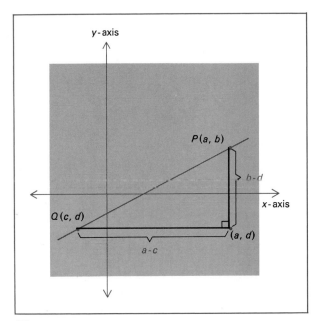

Figure 7.5

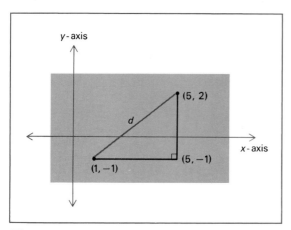

Figure 7.6

Solution: Construct a right triangle as shown in Figure 7.6. (Actually, there are two right triangles that could be constructed; either one can be used.) The coordinate of the third vertex is $(5, -1)$; the length of the vertical leg is $2 - (-1) = 3$; and the length of the horizontal leg is $5 - 1 = 4$. According to the Pythagorean theorem, if the length between the given points is denoted by d, then

$$d^2 = 3^2 + 4^2.$$

Hence $d^2 = 25$ and so $d = 5$.

Exercises 7.1

1. Find the distance between the points
   (a) $(0, 3)$ and $(4, -3)$.
   (b) $(-1, 3)$ and $(-3, 2)$.
   (c) (a, b) and (c, d), where $a < c$ and $b < d$.
2. What can you say about the coordinate (x, y) of a point P that lies above the x-axis? Below the x-axis? To the left of the y-axis? To the right of the y-axis?
3. Three of the vertices of a square are located at the points $(3, 1)$, $(3, -3)$, and $(-1, -3)$. Find the coordinate of the fourth vertex and the coordinate of the center of the square.
4. The base of an isosceles triangle has endpoints $(-2, 2)$ and $(4, 2)$. The height of the triangle is 5. What is the coordinate of the third vertex?
5. Draw the circle whose center is at the point $(3, 1)$ and that passes through the point $(7, -2)$. What is the radius of this circle?
6. A triangle has its vertices at the points $(-2, 2)$, $(7, 2)$, and $(2, -2)$. Is this triangle a right triangle or not?

7. Prove that the points $(-1, 2)$, $(4, -3)$, and $(5, 3)$ are the vertices of an isosceles triangle.
8. Do the points $(8, 5)$, $(-4, -11)$, and $(-6, 2)$ all lie on the same circle with center at $(2, -3)$?

7.2 Lines and Linear Equations

We have observed the connection between the geometric objects that are points and the arithmetic objects that are ordered pairs of real numbers. In this section we shall discuss the connection between the geometric objects that are lines and the arithmetic objects that are equations of a certain type.

Consider the line shown in Figure 7.7. We have identified a few points of this line by giving their coordinates. Is there anything special or unusual about these coordinates? Do they display any kind of pattern? If the x- and y-coordinates are added, the sum is 4. That is, if (x, y) is one of these points, then $x + y = 4$. This may not appear startling, but see what has happened: We began with a geometric object (the line) and have produced an algebraic object (the equation $x + y = 4$). Here is another example of how we can produce an equation from a line.

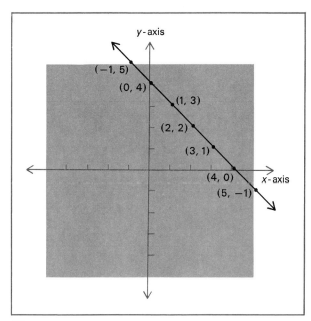

Figure 7.7

Example 1. In Figure 7.8 we have identified a line and some of the points of that line. The special relationship between the x- and y-coordinates of these points is not so obvious, but a careful examination will reveal this relationship. It is that the x-coordinate made smaller by 1 is equal to twice the y-coordinate. Thus, if we denote a typical point of the line by (x, y), then we may conclude that $x - 1 = 2y$. Once again a line has given rise to an equation.

Let us examine the equations we obtained in these two examples. The first equation was

$$x + y = 4$$

and the second equation (rewritten) was

$$x - 2y = 1.$$

These equations have the form of what is called a linear equation. **A linear equation** is one that can be rewritten in the general form

$$ax + by = c,$$

where a, b, and c are real numbers and not both a and b are zero. Other examples of linear equations are $3x - 2y = -5, y = 5 - 2x, y = 2x, 2x = 5,$ and $y = 1$. The

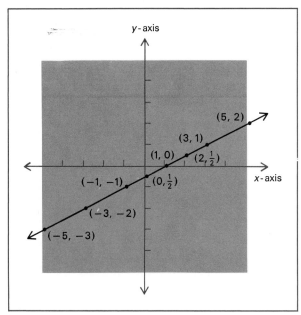

Figure 7.8

preceding two examples would seem to suggest that each line in the plane gives rise to a linear equation, and this is indeed true. By examining the coordinates of the points of a line, we can always find a linear equation. In these two examples we found the equations by inspection. There is a routine procedure for finding the equation of a line; however, for the time being, the inspection method will suffice.

By a **solution** of a linear equation we mean numerical values of the variables x and y that produce a true statement. For example, $x = 2$ and $y = 1$ together comprise a solution of the linear equation $x - y = 1$ because $(2) - (1) = 1$ is a true statement. We write this solution in the form of an ordered pair: $(2, 1)$. The left-hand part of the ordered pair is the x value of the solution and the right-hand part is the y value of the solution. The ordered pair $(4, 3)$ is another solution of $x - y = 1$, since $(4) - (3) = 1$ is a true statement. It should be clear that linear equations in two variables have infinitely many solutions. Among the solutions of the equation $x - y = 1$ are the solutions $(2, 1)$, $(3, 2)$, $(4, 3)$, $(5, 4)$, and so on.

Example 2.   Some solutions of the linear equation $x - 2y = 1$ are $(0, -\frac{1}{2})$, $(2, \frac{1}{2})$, $(3, 1)$, $(-3, -2)$, and $(18, \frac{17}{2})$. These solutions were found by giving various values to x and then determining the corresponding values of y. Of course, we can just as well give values to y and then compute the corresponding values of x. For instance, when $y = 0$, we get the equation $x - 2(0) = 1$ so that the ordered pair $(1, 0)$ is also a solution.

Example 3.   Find five solutions of the linear equation $2x = 4$.

Solution: If we first rewrite this equation in the form $ax + by = c$, it will be clearer what the solutions are. When we rewrite $2x = 4$ in this form, we get

$$2x + 0y = 4.$$

Now it should be clear that any ordered pair $(2, y)$ (where y is any number at all) is a solution. Five solutions are $(2, 1)$, $(2, 2)$, $(2, 3)$, $(2, 4)$, and $(2, 5)$.

We have seen that lines give rise to linear equations. The reverse is true as well. That is, if the solutions of a linear equation are located in the coordinate plane, then these solutions form a straight line. For example, consider the linear equation $2x + y = 2$. A few solutions of this equation are $(0, 2)$, $(1, 0)$, $(-1, 4)$, $(2, -2)$, $(-2, 6)$, and $(3, -4)$. We have located the points named by these ordered pairs in Figure 7.9. The reader will not have to stretch his imagination very far to conclude that these points do in fact belong to the same straight line, and that if we were to identify more solutions of the linear equation, then these solutions would also belong to this same line. We call the line that is formed by the solutions of a linear equation the **graph** of the equation. Thus the line shown in Figure 7.9 is the graph of the linear equation $2x + y = 2$.

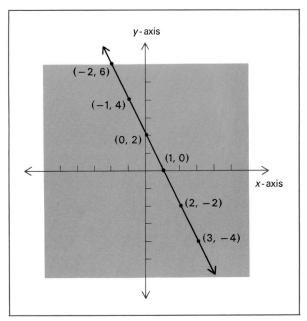

Figure 7.9

Example 4. Find the graph of the equation $x - 2y = 3$.

Solution: Since a line is determined by any two of its points, all we need to do is to find two solutions of this equation and locate these in the plane. The line determined by these two points is the graph of the given equation. The easiest solutions to find are those in which either x or y is 0. If $x = 0, y = -\frac{3}{2}$; thus $(0, -\frac{3}{2})$ is a solution. If $y = 0$, $x = 3$; thus $(3, 0)$ is a solution. The unique line passing through these two points is the graph of the linear equation (see Figure 7.10).

The two points found in Example 4 are called the **intercepts** of the line. The point $(0, -\frac{3}{2})$ is called the ***y*-intercept** of the line and $(3, 0)$ is called the ***x*-intercept** of the line.

We have seen that there is an intimate connection between lines and linear equations. The connection between a line and its equation is that the set of coordinates of the points of the line is equal to the set of solutions of the linear equation. This connection may be used to the mutual advantage of both geometry and algebra. However, in practice, the invention of analytic geometry has resulted in the subversion of geometry to the advantage of algebra. Problems that prior to the introduction of analytic geometry would have been solved geometrically were solved algebraically after the introduction of analytic geometry. There are a number of reasons for this, but perhaps the most significant one is that geometric arguments tend to stress inventiveness and cleverness more than algebraic argu-

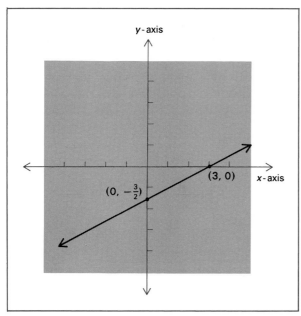

Figure 7.10

ments. On the whole, algebraic arguments are more routine, and what must be done to complete one is more self-evident than is the case with geometric arguments.

Exercises 7.2

1. Find the equation of the line that passes through the following points.
   (a) $(1, -1)$, $(3, 1)$, and $(5, 3)$.
   (b) $(-1, 4)$, $(0, 2)$, $(1, 0)$, $(2, -2)$, and $(3, -4)$.
2. Write each of these linear equations in the form $ax + by = c$.
   (a) $3 - 2y = 4x$.   (b) $6x = 17$.   (c) $3y + 6 = -2x$.
3. Find two solutions of each equation, and use these solutions to graph the line corresponding to each equation.
   (a) $5x - 3y = -1$.   (b) $3x + 2y = 4$.
4. Graph these equations by locating their x- and y-intercepts.
   (a) $4x - 2y = 1$.   (b) $6x = 14$.   (c) $3x - 7y = -4$.
5. What can you say about the graph of an equation each of whose solutions has the same x-coordinate? Each of whose solutions has the same y-coordinate?
6. What can you say about the graph of the equation $ax + by = 0$?
7. Graph the equations $2x - y = 3$ and $4x - 2y = 6$. Can you find other equations that have the same line for their graph? Given a line, how many equations are there that have that line for their graph?

7.3 The Slope of a Line

We know that a line is completely determined as soon as two of its points are given. Certainly, one point alone does not determine a line. Thus to say "Consider the line determined by the point $(1, 2)$" is nonsense because there are infinitely many different lines passing through the point $(1, 2)$. But, if we identify one of the line's points and also give some indication as to the slope of that line, then the line is determined. For example, we could say, "Consider the line passing through the point $(1, 2)$ and making an angle of $45°$ with the x-axis." This does describe a unique line, the line shown in Figure 7.11. (Incidentally, angles are usually measured in the counterclockwise direction.) Therefore, one way to describe the slope of a line is to describe the angle that that line makes with the x-axis. But there is another way to describe the slope that is more useful because it relates directly to the equation of the line. We shall study this method of describing the slope of a line in this section.

A common way to measure the degree of incline or decline of a highway is to measure the amount of vertical rise or fall over a certain horizontal distance. For example, a highway that rises vertically 10 feet over a horizontal distance of 1000 feet would be said to have slope of $\frac{10}{1000}$ or $\frac{1}{100}$. If the highway fell 10 feet vertically for every 1000 feet horizontally, then the highway would have slope

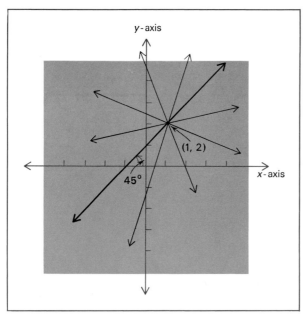

Figure 7.11

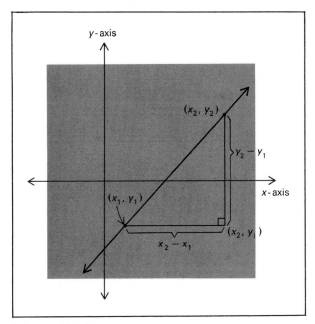

Figure 7.12

$-\frac{10}{1000}$ or $-\frac{1}{100}$. We use this same idea to measure the slope of a line. To determine the slope of a line, select any two distinct points (it does not matter which two points you choose), say, the points (x_1, y_1) and (x_2, y_2). The slope of the line (refer to Figure 7.12) is then defined to be the quotient

$$\frac{y_2 - y_1}{x_2 - x_1}.$$

Here are some examples of how the slope of a line is computed.

Example 1. The line m in Figure 7.13 passes through the points $(1, 2)$ and $(3, 5)$ and so has slope

$$\frac{5 - 2}{3 - 1} = \frac{3}{2}.$$

The slope of line ℓ in Figure 7.13 is

$$\frac{3 - (-1)}{(-4) - (-2)} = \frac{4}{-2} = -2.$$

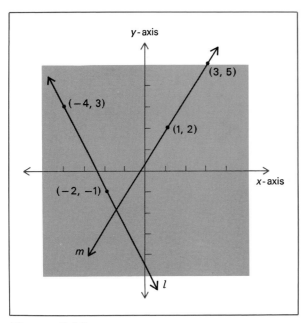

Figure 7.13

Example 2. Draw a picture of a line that is parallel to the x-axis. If (x_1, y_1) and (x_2, y_2) are points of this line, then, since $y_1 = y_2$, the expression for the slope is $0/(x_2 - x_1) = 0$. Hence a line parallel to the x-axis has slope of zero.

Example 3. If a line is parallel to the y-axis, and if (x_1, y_1) and (x_2, y_2) are two points of the line, then, since $x_1 = x_2$, the expression for the slope of the line is $(y_2 - y_1)/0$. This symbol is meaningless, and so lines parallel to the y-axis have no slope at all.

Examples 2 and 3 show that to say a line has zero slope is very different from saying that the line has no slope.

If a line is not parallel to the y-axis (that is, is not vertical), then its equation can be written in the special form

$$y = mx + b,$$

where m and b are real numbers. Now, if $x = 0$, then $y = b$; thus $(0, b)$ is the y-intercept of the line. In addition, the number m is the slope of the line. We can see this as follows: Suppose that (x_1, y_1) and (x_2, y_2) are points of this line. Then these pairs are solutions of the linear equation $y = mx + b$, and so we can write the equations

$$y_2 = mx_2 + b \quad \text{and} \quad y_1 = mx_1 + b.$$

By subtracting the second equation from the first, we can obtain

$$y_2 - y_1 = (mx_2 + b) - (mx_1 + b)$$
$$= mx_2 - mx_1$$
$$= m(x_2 - x_1).$$

Consequently, $m = (y_2 - y_1)/(x_2 - x_1)$ and m is the slope of the line. This means that we can identify the slope and the y-intercept of a nonvertical line simply by writing its equation in the form $y = mx + b$. For this reason, we call this form of the equation of a line the **slope-intercept form** of that equation.

Example 4.   Find the slope and y-intercept of the line $2x - 3y = 6$.

Solution: Writing this equation in the slope-intercept form, we get $y = \frac{2}{3}x - 2$; thus, the line has slope $\frac{2}{3}$ and y-intercept $(0, -2)$. The line is shown in Figure 7.14.

Example 5.   A line has slope $\frac{3}{2}$ and passes through the point $(1, 2)$. Find its equation.

Solution: We know that the equation of the line has the form

$$y = \frac{3}{2}x + b,$$

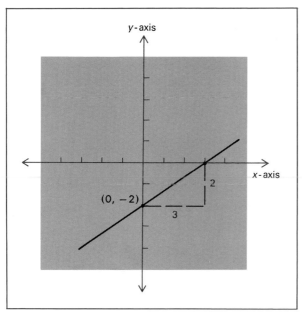

Figure 7.14

since the slope was given to us. It remains only to determine the number b. But since the line is known to pass through the point $(1, 2)$, we know that $(1, 2)$ is a solution of the equation we are looking for.

$$2 = (\tfrac{3}{2} \cdot 1) + b.$$

From this equation we see that $b = 2 - \tfrac{3}{2} = \tfrac{1}{2}$. Hence the equation of the line is

$$y = \tfrac{3}{2}x + \tfrac{1}{2}$$

or $3x - 2y = -1$.

Parallel lines must have the same slopes; therefore, we can tell whether or not two lines are parallel by examining their equations.

Example 6. Which of the lines $x - y = -1$ and $3x - y = 6$ is parallel to the line $x - y = 2$?

Solution: Rewriting these equations in slope-intercept form, we are asking which of the lines $y = x + 1$ and $y = 3x - 6$ is parallel to the line $y = x - 2$. The line $y = x + 1$ has the same slope as the line $y = x - 2$ (both have slope 1) and so these lines are parallel. The line $y = 3x - 6$ has slope 3 and is not parallel to $y = x - 2$ (see Figure 7.15).

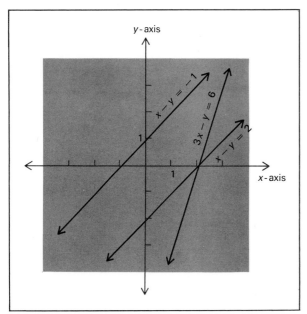

Figure 7.15

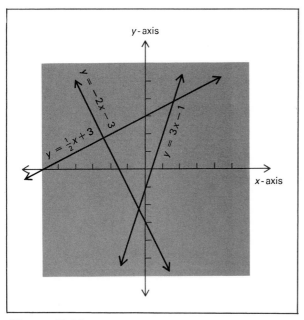

Figure 7.16

We can also tell whether or not two lines are perpendicular by looking at their equations. It can be proved that two lines are perpendicular if and only if the product of their slopes is equal to -1.

Example 7. Which of the lines $y = -2x - 3$ and $y = 3x - 1$ is perpendicular to the line $y = \frac{1}{2}x + 3$?

Solution: The line $y = 3x - 1$ is not perpendicular to $y = \frac{1}{2}x + 3$, since the product of the two slopes is not equal to -1. But $y = -2x - 3$ is perpendicular to $y = \frac{1}{2}x + 3$, since the product of their slopes (-2 and $\frac{1}{2}$) is equal to -1. These lines are shown in Figure 7.16.

Exercises 7.3 _____

1. A line passes through the points given. What is the slope of the line?
   (a) $(2, 2)$ and $(4, -2)$.   (b) $(-1, 2)$ and $(-4, 0)$.
2. Determine the slope of each line by locating two points belonging to the line.
   (a) $2x + y = 7$.   (b) $2x - y + 2 = 0$.
3. What can you say about a line that has no slope? What can you say about a line that has slope equal to 0? If a line has positive slope, then does the

line rise or fall as you move along the line from left to right? Does the line rise or fall if its slope is negative?

4. A line passes through the point $(-1, 3)$ and has slope $-\frac{1}{2}$. Graph the line and find its equation.

5. An isosceles right triangle has its legs parallel to the coordinate axes. What can you say about the slope of its hypotenuse?

6. Find three linear equations whose graphs form a right triangle.

7. Describe how you could use the idea of slope to prove that three given points were collinear. Test your idea by showing that the points $(-1, -5)$, $(2, 1)$, and $(3, 3)$ are collinear.

8. The vertices of a triangle are A, $(-4, 2)$; B, $(6, 6)$; and C, $(-3, -4)$. Find the equation of the altitude (that is, the perpendicular) from the vertex A to the side $\overline{BC}$.

9. In 1965 the population of Foosland, Illinois, was 13. In 1967 the population was 24. If you assume that the town's rate of growth is linear, what would you predict the population to be in 1971?

10. The height of a boy at age 6 was 1 meter and 10 centimeters. On his eighteenth birthday he was 2 meters, 30 centimeters in height. If this boy's growth were linear, how tall would you expect him to be when he is 24? Is human growth linear?

11. Suppose you put $1000 in a savings account and left it there for 30 years. Do you think that the growth of this money would be linear with respect to time? (Assume that interest is compounded daily.) If not, would it be faster than linear growth or slower than linear growth?

12. The thickness of the layer of barnacles on the underside of a ship's hull is linearly related to time. That is, if the thickness of the layer of barnacles is $\frac{1}{4}$ inch after two months, then it will be $\frac{1}{2}$ inch after four months, $\frac{3}{4}$ inch after six months, and so on. Can you think of other quantities besides thickness of barnacles and time that are related by a linear relationship?

7.4 The Conic Sections, I

The study of lines and linear equations is only the first part of the study of analytic geometry. The next curves of interest are called "the conic sections." These curves were studied by the Greeks, but their treatment using algebra had to await the invention of the coordinate plane. In this section we shall describe these curves without reference to the coordinate plane. In the next section we shall discuss them using the methods of analytic geometry.

A (right-circular) **cone** is a spatial surface, as shown in Figure 7.17. The cone may be thought of as consisting of the totality of all lines in space that pass through a fixed point P called the **vertex** of the cone and which make a fixed angle α

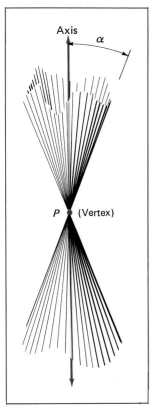

Axis

α

P (Vertex)

Figure 7.17

with a fixed line that passes through this vertex and is called the **axis** of the cone. A cone is a surface, not a solid, and extends without limit in two directions.

By cutting through this cone with planes, we form plane curves called sections of the cone or **conic sections.** If the cutting plane is held so that it is perpendicular to the axis of the cone, then the section so formed is a **circle** [see Figure 7.18(a)]. Thus a circle is a conic section. If the cutting plane is held as shown in Figure 7.18(b), then the resulting section is a sort of elongated circle called an **ellipse.** Holding the cutting plane as shown in Figure 7.18(c) produces a curve that is not closed and extends without limit in two directions. This curve is called a **parabola.** The fourth conic section is formed by holding the cutting plane so that it is parallel to the axis of the cone. The curve produced in this case [Figure 7.18(d)] consists of two branches, each of which somewhat resembles a parabola. This two-piece curve is called a **hyperbola.**

These definitions in terms of cuts of the cone are purely geometric in nature. We want to apply the methods of analytic geometry to these curves, however, and so would prefer to have definitions that may more easily be translated into algebraic language. We already have such a definition of the circle. We know that

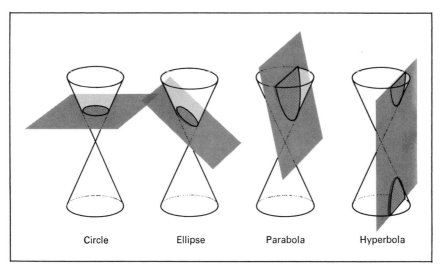

Circle Ellipse Parabola Hyperbola

Figure 7.18

a circle consists of all points in the plane that are a given distance away from a given point. We shall use this alternate definition of the circle in the next section to find an equation for the circle.

The other conic sections also have alternate definitions similar to the alternate definition given for a circle. An ellipse can be defined as the set of all points in the plane, the sum of whose distances from two given points is equal to a given distance greater than the distance between the two given points. Refer to Figure 7.19. In this illustration the given points are P and Q and the given distance is

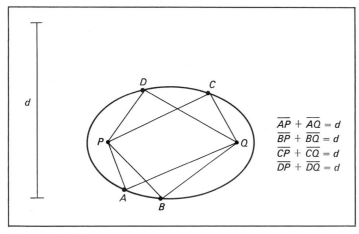

$$\overline{AP} + \overline{AQ} = d$$
$$\overline{BP} + \overline{BQ} = d$$
$$\overline{CP} + \overline{CQ} = d$$
$$\overline{DP} + \overline{DQ} = d$$

Figure 7.19

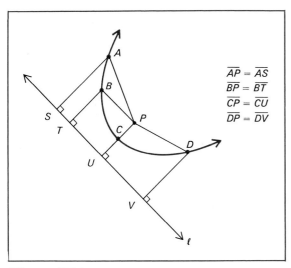

$$\overline{AP} = \overline{AS}$$
$$\overline{BP} = \overline{BT}$$
$$\overline{CP} = \overline{CU}$$
$$\overline{DP} = \overline{DV}$$

Figure 7.20

shown on the left and is called *d*. Notice that *d* is greater than the distance between *P* and *Q*. Each of the points *A*, *B*, *C*, and *D* is a point of the ellipse, since each of these points has the property that its distance from *P* plus its distance from *Q* is equal to the distance *d*. The points *P* and *Q* are called the **foci** of the ellipse.

A parabola is defined in terms of a given line and a given point. A parabola consists of all points in the plane whose distances from a given point are equal to their distances from a given line. Refer to Figure 7.20. The given point is *P* and the given line is *ℓ*. Each point of the parabola has the same distance from *P* as it does from the line *ℓ*. The point *P* is called the **focus** of the parabola, and the line *ℓ* is called the **directrix.**

A hyperbola (like an ellipse) is defined in terms of two points and a given distance. A hyperbola consists of all points in the plane, the difference of whose distances from two given points is equal to a given distance less than the distance between the two given points. In Figure 7.21, *P* and *Q* are the given points and *d* is the given distance. Each point of the hyperbola (which consists of two separate branches) has the property that the difference of its distances from *P* and from *Q* is equal to *d*.

The conic sections have innumerable important physical properties that make knowledge of them essential to an understanding of physics, astronomy, and many other sciences. Let us mention just a few of these.

1. If you wanted to enclose 1 acre of grazing land with a fence as cheaply as possible, you should make the fence circular. This is because a circle encompasses a given area with the least perimeter.
2. The reflectors of automobile headlights are parabolic in cross section

---

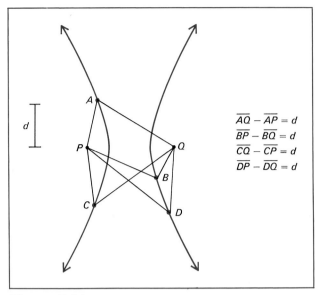

$$\overline{AQ} - \overline{AP} = d$$
$$\overline{BP} - \overline{BQ} = d$$
$$\overline{CQ} - \overline{CP} = d$$
$$\overline{DP} - \overline{DQ} = d$$

Figure 7.21

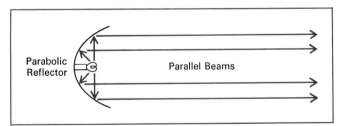

Figure 7.22

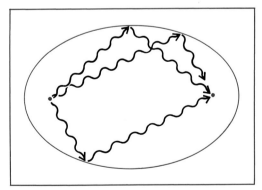

Figure 7.23

because, when the light source is located at the focus, the parabola reflects this light in parallel beams, thus maximizing the lighting effect (Figure 7.22).

3. If one person stands at one focus of an elliptically shaped room and whispers very softly, he will be heard by another person standing at the other focus because of a reflecting property of ellipses. Sounds emitted from one focus are reflected by the ellipse to the other focus (Figure 7.23).

4. The planets travel around the sun on elliptical paths. In each case the sun is located at one of the foci of these paths.

5. Meteors travel on paths that are conic sections, mainly ellipses.

Exercises 7.4

1. Can you think of a way to generate a cone by using rigid motions in space?

2. If we think of the light beams emitted by a flashlight as forming half of a cone, then the conic sections can be reproduced by shining a flashlight against a wall. For example, a circle can be produced by holding the flashlight at right angles to the wall. How would the other conic sections be demonstrated in this way?

3. Press two thumb tacks into a flat surface and attach a piece of string to the tacks. (Make the length of the string about double the distance between the tacks.) Now place a pencil so as to stretch the string and make it taut. Move the pencil about the surface in such a way as to keep the string taut at all times. The curve traced out is what kind of conic section?

4. How would you trace out a circle using string, tacks, and a pencil?

5. Draw a line on a piece of paper and locate a point 2 inches away from the line. Now locate 10 points whose distances from the line equal their distances from the point. Connect these points by a smooth curve. What kind of a conic section have you constructed?

6. Locate two points P and Q on a piece of paper so that they are 3 inches apart. Now locate a number of points that have the property that their distance from P less their distance from Q is equal to 1 inch. Locate a number of points with the property that their distance from Q less their distance from P is 1 inch. By connecting these points with smooth curves, you will have constructed a hyperbola.

7. The four conic sections we have described are called **nondegenerate conics.** The degenerate conics are the point, the line, and the figure consisting of two intersecting lines. Explain how these degenerate conics can be obtained by cutting a cone with a plane. The point is a degenerate case of what kinds of conic sections? The line is a degenerate case of which conic? The pair of intersecting lines is a degenerate case of what kind of conic?

8. *Pascal's Theorem* states that if six points A, B, C, D, E, and F are chosen on a conic section, and if the lines $\overleftrightarrow{AB}$, $\overleftrightarrow{BC}$, $\overleftrightarrow{CD}$, $\overleftrightarrow{DE}$, $\overleftrightarrow{EF}$, and $\overleftrightarrow{FA}$ are drawn, then the intersections of the lines $\overleftrightarrow{AB}$ and $\overleftrightarrow{DE}$, $\overleftrightarrow{BC}$ and $\overleftrightarrow{EF}$, and $\overleftrightarrow{CD}$ and $\overleftrightarrow{FA}$ are collinear. Illustrate this using a circle. Repeat the illustration using a parabola.

7.5 The Conic Sections, II

We have seen how the conic sections can be defined in terms of cuts of a cone and how they can be defined in terms of points, lines, and distances. These latter definitions lend themselves to the techniques of analytic geometry because points are easily changed into ordered pairs, lines into linear equations, and distances into numbers. In this section we shall demonstrate how equations can be found for some of the conic sections using the point-line-distance definitions given in Section 7.4.

Let us begin by finding the equation of the circle whose center is at the point (a, b) and whose radius is r units. Refer to Figure 7.24. If we let (x, y) be the coordinate of a typical point P on this circle, then, according to the definition of a circle, the distance between (x, y) and (a, b) is equal to r. Therefore, we are able to determine the lengths of the three sides of the right triangle shown in Figure 7.24. Using the Pythagorean Theorem, we may then write the equation

$$(x - a)^2 + (y - b)^2 = r^2.$$

We have found an equation in two variables, the solutions of which are the coordinates of the points belonging to this circle. Conversely, every point belonging to this circle has a coordinate that is a solution of this equation. Just as we were able to establish a connection between straight lines and linear equations, now

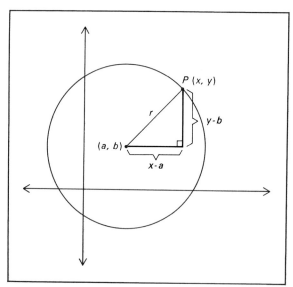

Figure 7.24

we have been able to establish a connection between circles and equations of the form $(x - a)^2 + (y - b)^2 = r^2$, where a, b, and r are real numbers.

Example 1. What is the equation of the circle whose center is at the origin and which has a radius of 4 units?

Solution: We use the equation $(x - a)^2 + (y - b)^2 = r^2$ with $a = 0, b = 0$, and $r = 4$. We get $(x - 0)^2 + (y - 0)^2 = 4^2$ or $x^2 + y^2 = 16$.

Example 2. The equation $(x + 4)^2 + (y - 1)^2 = 25$ is the equation of a circle. Find the center and radius of the circle.

Solution: To find the center and radius, we must rewrite this equation in the form $(x - a)^2 + (y - b)^2 = r^2$. When we do so, we get $[x - (-4)]^2 + (y - 1)^2 = (5)^2$ so that the circle has its center at $(-4, 1)$ and has a radius of 5 units.

We shall not derive the equations of ellipses or hyperbolas because to do so requires more algebraic manipulation than we wish to become involved with. But working with parabolas is not too much, so let us consider the problem of finding the equation of a parabola. Recall that a parabola is the set of all points that are equidistant from a fixed point and a fixed line. For example, if the fixed point is the point $(0, 2)$ and the fixed line is the x-axis (the line $y = 0$), then there is defined a parabola consisting of all the points of the plane whose distance from $(0, 2)$ is equal to their distance from $y = 0$. We let (x, y) be the coordinate of a typical point P of this parabola and compute the lengths of the segments $\overline{PA}$ and $\overline{PB}$, as shown in Figure 7.25.

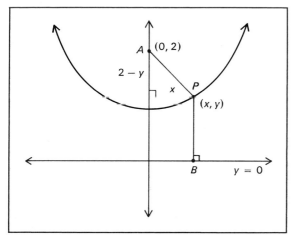

Figure 7.25

$$(\overline{PA})^2 = (2 - y)^2 + x^2$$
$$\overline{PB} = y.$$

By definition of the parabola, the point (x, y) belongs to the parabola if and only if $\overline{PA} = \overline{PB}$. Thus

$$\sqrt{(2 - y)^2 + x^2} = y,$$

or, upon squaring both sides of this equation,

$$(2 - y)^2 + x^2 = y^2.$$

This is the equation of the parabola. A point belongs to the parabola if and only if the coordinate of that point is a solution of this equation. We would usually want to simplify this equation like this:

$$(4 - 4y + y^2) + x^2 = y^2$$
$$4 - 4y + x^2 = 0.$$

All of the equations we have seen in this section are of the same general form. They are all **equations of degree 2,** by which we mean an equation that can be rewritten in the special form

$$ax^2 + bxy + cy^2 + dx + ey + f = 0,$$

where a, b, c, d, e, and f are real numbers and not all of a, b, and c are zero. (If a, b, and c were zero, then this equation would be a linear equation.) It can be proved that every conic section has an equation that is an equation of degree 2. We shall not give details, but a great deal of information about the conic sections can be obtained from their equations. Indeed, all of the geometric information about these conic sections is contained one way or another in their equations.

Finally, let us use the following examples to illustrate the way that plotting points can help in graphing conic sections.

Example 3. The equation $4x^2 + y^2 = 16$ has degree 2, and so its graph is a conic section. In this case the conic section is an ellipse. To sketch this ellipse (since we know what its general shape is), we need only identify a few solutions of the equation, locate these solutions in the coordinate plane, and then connect these points together with a smooth curve. Some solutions are presented in table form,[1] and the graph is sketched in Figure 7.26.

---

[1] This table is a convenient listing of some of the ordered pairs that are solutions of the given equation and conveys the information that the following ordered pairs are solutions of the equation: $(0, 4)$, $(0, -4)$, $(0, 0)$, $(-2, 0)$, $(1, 3.4)$, $(1, -3.4)$, $(-1, 3.4)$, and $(-1, -3.4)$. We are using 3.4 as an approximate value for $\sqrt{15}$.

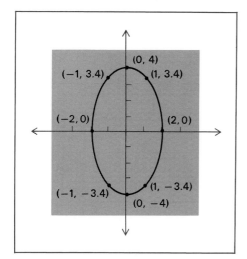

x	y
0	4, −4
2	0
−2	0
1	3.4, −3.4
−1	3.4, −3.4

Figure 7.26

Example 4. Sketch the graph of the parabola whose equation is $y = x^2$.

Solution: We know what the general shape of a parabola is, so we need only locate enough points of this parabola to be able to determine its particular shape. In Figure 7.27 we have tabulated some points and sketched the parabola.

Example 5. Sketch the hyperbola that is the graph of the equation $x^2 - y^2 = 4$.

Solution: See Figure 7.28. Do not forget that a hyperbola consists of two branches.

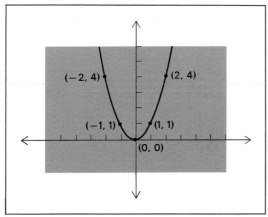

x	y
0	0
1	1
−1	1
2	4
−2	4

Figure 7.27

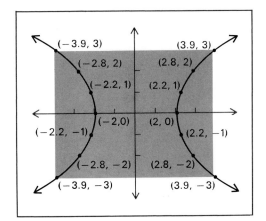

x	y
0	No values
± 2	0
$\pm \sqrt{5} \doteq 2.2$	± 1
$\pm \sqrt{8} \doteq 2.8$	± 2
$\pm \sqrt{13} \doteq 3.9$	± 3

Figure 7.28

Exercises 7.5

1. Write the equation of the circle with the given point as center and the given distance as radius.

   (a) $(1, 5)$; 6.　　　(b) $(-3, -5)$; distance between $(1, 8)$ and $(-2, 5)$.

2. Identify the center and radius of each of these circles.

   (a) $(x + 1)^2 + (y - 1)^2 = 16$.　　　(b) $(x - 5)^2 + (y + 4)^2 = 10$.

3. Find the equation of the parabola defined by the line $x = 0$ and the point $(2, 3)$.

4. Graph these equations.

   (a) $x^2 - y^2 = 4$ (hyperbola).　　　(b) $9x^2 + 25y^2 = 225$ (ellipse).

   (c) $y^2 = 8x$ (parabola).　　　(d) $y^2 = -8x$.

   (e) $4x^2 + 3y^2 = 12$.　　　(f) $16y^2 - 9x^2 = 144$.

   (g) $x^2 = 8y$.　　　(h) $x^2 = -8y$.

5. What can you say about the general shape of the parabola $x^2 = ay$ if $a > 0$? What if $a < 0$? What is the general shape of the parabola $y^2 = ax$ if $a > 0$? What if $a < 0$? (*Hint:* Consider various parts of Exercise 4.)

Functions

The word "function" has a number of different meanings outside of mathematics, but its meaning in mathematics is exact. We have been involved with functions from the beginning of our study, but it has not been necessary to specifically point this out. Now is a convenient time to discuss the concept, since this discussion is made easier once the Cartesian coordinate plane is available. The study of functions comprises one of the largest and most important branches of mathematics, and the idea pervades every part of mathematics. The power of the concept lies with its almost total generality, but in our work we shall be dealing mainly with a kind of function called a "real-valued function."

8.1 Black Boxes

We shall introduce the concept of function by means of a physical analogy called "the black box." All of the important ideas related to the concept of function can be illustrated using this analogy.

A **black box** is a mysterious-looking device something like the object shown in Figure 8.1. This box has an opening in its top called the "input opening" and a spout near its bottom called the "output spout." One puts things into the top

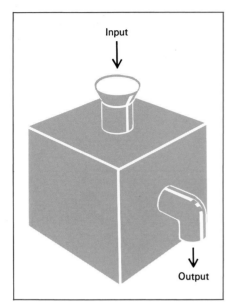

Input

Output

Figure 8.1

and things come out of the bottom. What comes out depends upon what was put into the box and upon the particular internal construction of the black box. The inputs are the objects that are put into the box at the top, and the outputs are the objects that come out of the box at the bottom.

Example 1. There is a black box that uses whole numbers as inputs and produces whole numbers as outputs. This box is constructed in such a way that if you put the number 1 into the box, the number 4 drops out the output spout. If you put 16 into the box, 19 comes out the bottom. The input 234 results in the output 237. Perhaps it is clear from these examples that this particular black box works by adding 3 to any inputs that are inserted into the box. We might call this particular box the "add 3" black box.

The black box discussed in Example 1 has a property common to all properly constructed black boxes. Given one input, there is produced by the box *exactly one* output. Put one object into the box and exactly one object comes out the bottom. This is an especially important property of a black box.

Example 2. In 1532 a German alchemist named O. Zeltmacher invented a black box that worked as follows: You would take a piece of paper and draw on it the picture of a polygon. Then you would put this paper into the input opening. After some whirrings, out from the box would drop another piece of paper on which was written the number of vertices of the polygon you had inserted into the box.

If you put in a picture of a square, the output would be a piece of paper with the number "four" written on it. This box took polygons as input and gave out whole numbers as output.

Zeltmacher also tried to invent a black box that would work the reverse of his successful "vertex counting" box. He tried to design a box that would take as input pieces of paper with whole numbers greater than 3 written on them. The output was supposed to be a piece of paper with a picture of a polygon having as many vertices as the input number. To test the box, Zeltmacher inserted a piece of paper with the word "three" written on it. The box began to pour out all sorts of triangles. In trying to produce all of the infinitely many different triangles, the box overheated and blew up. That is all we hear of Zeltmacher. The trouble was that he tried to design a black box that gave more than one output for a given input.

Example 3. There is a black box that eats typewriters as input and gives out whole numbers as output. It might be called the "serial number" black box. If the typewriter on which this book was typed were put into this box, the number 61554409 would come out.

There need be no special relationship between the inputs and the outputs of a black box. These boxes (being after all quite imaginary) can be constructed in any way at all. They can be made so as to accept anything at all as inputs and to produce anything at all as outputs. Of course, in mathematics these black boxes usually "act" upon mathematical objects and "produce" mathematical objects. Here is an example of such a black box.

Example 4. The inputs for this black box are linear equations and the outputs are real numbers. Here are a few examples of how this particular black box works.

$$
\begin{array}{ccc}
\textit{Inputs} & \rightarrow & \textit{Outputs} \\
2x + y = 4 & \rightarrow & 2 \\
y + 4 = -9 & \rightarrow & 0 \\
y - 3x - 4 = 0 & \rightarrow & -3.
\end{array}
$$

Can you describe in words how this box works? It takes an input (a linear equation) and produces from it an output, which is the real number that is the coefficient of the x term of the given input.

The black box in Example 4 is a properly constructed box because, given one input, exactly one output is produced. You should be careful not to confuse this characteristic of a black box with the fact that different inputs may produce the same output. For example, this black box produces the same output, 2, from the two different inputs $2x + y = -1$ and $2x - 15y = 0$, but there is nothing wrong in this. Many different inputs may produce the same output, but one input may not produce many different outputs.

Each time we have described a black box, we have been careful to describe the kind of objects that could be used as inputs for that box. When describing a black box, we must not only describe the effect of the box upon a given input, but we must also specify the kind of objects that can be used as inputs.

Example 5.   There is a black box that can be called the "positive square root box." This black box operates upon an input number by producing the positive square root of that number. But note that this black box is legitimate only if the inputs are restricted to the set of nonnegative real numbers. For example, if we were to allow -1 to be an input for this box, then no output would be produced (since there is no real number that is the square root of -1). Hence we must specifically state that this black box operates only upon nonnegative inputs.

In the next section we shall replace the notion of a black box with its mathematical equivalent—the notion of a function.

Exercises 8.1 _____

1. There is a black box called the "negation box" that takes statements as inputs and gives out statements as outputs. What comes out of the box when these statements are put into it?
    (a) $2 + 2 = 5$.   (b) $2 + 3 = 6 - 1$.
    (c) $\sim p \vee \sim q$.
2. Describe the internal workings of a black box that can take the objects of the first set as inputs and give out the objects from the second set as outputs.
    (a) $\{1, 2, 3, 4, 5, 6, 7, 8\}$; $\{a, b, c, d, e, f, g, h\}$.
    (b) {bad, whole, ugly, evil, wicked, craven}; {incomplete, good, beautiful}.
    (c) {house, car, spigot, horseshoe, swatch}; {Man-O'-War, wrench, beer, needle, doormat}.
3. O. Zeltmacher tried to construct the following black boxes but they all blew up. Why?
    (a) A black box such that, when a simple closed curve was put into the box, another simple closed curve topologically equivalent to it would drop out the bottom.
    (b) A black box that, when fed a whole number, would produce a whole number less than the one put into the input spout.
    (c) A box such that, whenever a woman was put into the top, there would drop out at the bottom the only man she ever loved.
    (d) A box such that, when a letter was put into the input spout, a word containing that letter would drop from the output spout.
4. Is the black box that, when any real number is put into it, produces the number 4 a legitimate black box?
5. Unless we carefully describe the kinds of input objects that can be used for

the following black boxes, they will not always produce an output object. Describe the set of permissible inputs for each box.

(a) The box that, when given a line, produces its slope.

(b) The box that, when given a conic section, produces the area contained by that conic.

(c) The box that, when given the equations of two lines, produces the coordinates of their point of intersection.

8.2 The Concept of Function

We shall continue our introduction to the concept of function using mathematical terminology and notation.

The black boxes discussed in Section 8.1 are examples of functions. Each of these black boxes can be described as follows. A black box is a device by means of which, corresponding to each object in a certain set of input objects (this set depending upon the particular black box we are describing), exactly one output object is produced. We call the black box a **function,** the set of inputs the **domain** of the function, and the set of outputs the **range** of the function. To describe a function, therefore, it is necessary to specify two sets, the domain and the range, and then to describe a rule or instruction by means of which, corresponding to each and every object in the domain set, exactly one object in the range set can be determined. The idea of a function, then, involves three different ideas: the domain of the function, the range of the function, and the rule for performing the function. Usually, however, it is necessary to be concerned only about the domain and the rule—the range will take care of itself. That is, as soon as the domain and the rule are given, the range set is automatically determined.

Example 1. If the domain is the set $\{1, 2, 3\}$ and the rule for performing the function is "To perform the function upon a domain object, add 3 to that domain object," then the range of the function must be the set $\{1 + 3, 2 + 3, 3 + 3\}$ or $\{4, 5, 6\}$.

Example 2. The diagram

$$a \to 3$$
$$b \to 1$$
$$c \to 2$$

describes a function whose domain is the set $\{a, b, c\}$ and whose range is the set $\{1, 2, 3\}$. The arrows define the effect of the function upon each domain object.

Example 3.   Consider the function defined on the domain of all real numbers and whose rule is "To perform this function upon a domain object (that is, a real number), add 3 to that domain object." This function is different from the function in Example 1 because it has a different domain from that function. Many students mistakenly think that because these two functions are performed according to the same rule, they are the same function. Two functions can be the same only if their domains and their rules are the same. We call the function in this example an *extension* of the function in Example 1.

Whenever possible, we use equations to describe functions. For example, consider the function in Example 1. Denoting this function symbolically by the letter f (suggestive of the word "function"), we would symbolically describe the effect of this function upon the domain object 1 by writing

$$f(1) = 4.$$

The symbol $f(1)$ is read "f of 1." We would describe the effect of this function upon 2 and 3 by writing $f(2) = 5$ and $f(3) = 6$. Thus the three equations

$$f(1) = 4$$
$$f(2) = 5$$
$$f(3) = 6$$

completely describe this function. These equations tell us not only how the function works upon domain objects, but they also tell us what the domain objects are. (They tell us what the range objects are, too.) To describe the effect of the function generally, we would write

$$f(x) = x + 3, \quad \text{where } x \in \{1, 2, 3\}.$$

The function in Example 3 would be described using this notation by writing

$$f(x) = x + 3, \quad \text{where } x \text{ is a real number.}$$

Observe that we mention not only the rule, but also the domain. This notation is called *functional notation*. Here are more examples of functions described using this notation.

Example 4.   The equation $f(x) = 2x + 3$ (where x denotes an integer) describes a function that acts upon integers by doubling them and then adding 3 to the result. The range objects produced by this function are integers, so we say that the function f is a function from the integers into the integers.

Example 5.   If x is an integer, then the distance of that integer from the origin on the number line is called the **absolute value** of that integer. The absolute value of an integer x is denoted by the symbol $|x|$. For example, $|-6| = 6$ and $|9| = 9$. The function f defined according to the equation

$$f(x) = |x|, \quad \text{where } x \text{ is an integer,}$$

is a function whose domain is the set of all integers and whose range is the set of all nonnegative integers. For example, $f(-6) = 6$ and $f(9) = 9$. This function is called the **absolute value function.**

Example 6. If x is a real number, then the symbol $[\![x]\!]$ represents the greatest integer less than or equal to x. For example, $[\![1]\!] = 1$, $[\![\frac{2}{3}]\!] = 0$, $[\![\pi]\!] = 3$, and $[\![-19\frac{7}{8}]\!] = -20$. The function defined by the equation

$$f(x) = [\![x]\!], \quad \text{where } x \text{ is a real number,}$$

is a function whose domain consists of all real numbers and whose range is the set of all integers. This function is called the **greatest integer function.**

Not all functions can easily be described using equations, but all of the functions that we shall be working with can be. Also, most of our functions will have as their domains the set of all real numbers.

We remarked in the introduction to this chapter that we have been working with functions since the beginning of our study of mathematics. Here are two examples of functions that we have dealt with without specifically calling attention to the fact that we were doing so. The first is an example from arithmetic.

Example 7. Let f be the function whose domain consists of all *ordered pairs of integers,* which is described according to the equational rule

$$f((a, b)) = a + b.$$

This function acts upon ordered pairs of integers and produces integers according to the following rule: Add the left-hand and right-hand parts of the given ordered pair. The function acting upon the ordered pair $(2, 3)$ produces the integer $2 + 3$ or 5, and so we write $f((2, 3)) = 5$. Similarly, $f((-1, -5)) = -6$. This function is called the **addition of integers function.** To say that addition of integers is commutative means simply that this function acts the same way upon the pair (a, b) as it does upon the pair (b, a):

$$f((a, b)) = f((b, a)).$$

The operations of subtraction and multiplication of integers similarly can be regarded as functions from the domain of all ordered pairs of integers into the integers. The **subtraction of integers function** acts upon the ordered pair $(-2, -7)$

and produces the integer $2 - (-7)$ or 5. The **multiplication of integers function** acting upon this same ordered pair results in the integer 14.

Our next example comes from Euclidean geometry.

Example 8. Let the function f be defined on the set of all polygons as follows: If P is a polygon, then $f(P)$ is that real number which is the area of the polygon P. The domain objects for this function are polygons and the range objects are real numbers. Thus, for example,

$$f\left(2\ \boxed{}_{3}\right) = 6$$

and

$$f\left(5\ \triangle_{4}\right) = 10.$$

This function might be called the **area function.**

Exercises 8.2

1. In Example 8 we described a function whose domain consisted of all polygons and whose range was a subset of the real numbers. Describe another such function. Describe a function whose domain is the set of all circles and whose range is a subset of the set of all real numbers.
2. If f denotes the *absolute value function* with domain the integers, then what can you say about x if $f(x) = 5$?
3. If f denotes the *greatest integer function* with domain the real numbers, then what can you say about x if $f(x) = 2$?
4. Define a function whose domain is the set $\{a, b, c\}$ and whose range is a proper subset of the set $\{1, 2, 3\}$.
5. Let f denote the *subtraction of integers function*. This function has as its domain the set of all ordered pairs of integers and has as its range the set of integers. List ten ordered pairs (x, y) such that $f((x, y)) = 6$.
6. Let the *truth value function* be denoted by t. This function has as its domain the

set of all statements and as its range the set {true, false}. Complete the following:

(a) t(All men are mortal) $= ?$ (b) $t(2 + 2 = 4) = ?$

(c) $t(p \wedge \sim p) = ?$ (d) $t(p \vee \sim p) = ?$

(e) t(Some implications have true converses) $= ?$

(f) t(All implications have true converses) $= ?$

(g) t(All implications have true contrapositives) $= ?$

7. Let p denote the *perimeter function* with domain the set of all polygons.

  (a) Is 0 an object in the range of this function? Are there any negative numbers in the range of this function?

  (b) Give three examples of polygons such that p(polygon) $= 4$.

  (c) How many squares are there such that p(square) $= 16$?

  (d) How many rectangles are there such that p(rectangle) $= 16$?

  (e) Find a triangle and a rectangle such that p(triangle) $= p$(rectangle).

8. Define, using functional notation, functions that have as their domains the first of the given sets and have their ranges contained in the second of the given sets. (That is, the ranges need not be equal to the second sets.)

  (a) $\{0, 1, 2, 3\}$; $\{0, 1, 2, 3\}$.

  (b) $\{0, 2, 4, 6, 8, \ldots\}$; $\{0, 1, 2, 3, 4, 5, 6, \ldots\}$.

  (c) $\{$triangle, square, pentagon, hexagon$\}$; $\{0, 1, 2, 3, 4, 5, 6, 7, 8\}$.

  (d) $\{\frac{1}{2}, \frac{2}{3}, \frac{5}{3}, 17, \frac{21}{5}, 0\}$; $\{0, 4, 17, 1, 189, 32\}$.

8.3   Graphing Functions

It is particularly useful to visualize functions geometrically as curves in the coordinate plane. In fact, we did some of this in the last chapter when we graphed certain equations. Information about the function itself can be obtained from its geometric "picture."

Let us begin by examining the function $f(x) = 3x - 7$ (where x represents any real number). We can graph this function by constructing a coordinate system (just as in the last chapter) with one axis labeled the x-axis and the other labeled the $f(x)$-axis. Then we construct a table and compute a few values of $f(x)$:

x	-1	0	1	2	3
$f(x)$	-10	-7	-4	-1	2

We now know the coordinates of five points of the graph of this function. These points are shown in Figure 8.2. This function is called a **linear function.** In Chapter 7 we wrote this function as $y = 3x - 7$ instead of $f(x) = 3x - 7$.

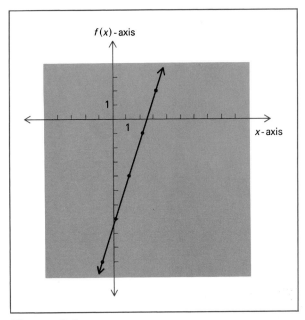

Figure 8.2

Example 1. Graph the function defined on the domain of all real numbers according to the equational rule $f(x) = 6x^2 - 1$. Plot points corresponding to every integer value of x between $x = -3$ and $x = 3$.

Solution: We first construct a table of values:

x	-3	-2	-1	0	1	2	3
$f(x)$	53	23	5	-1	5	23	53

Then we plot the points we have found, as shown in Figure 8.3. These points, if properly connected, will give us a rough idea as to the graph of the function. There are many possibilities for making a mistake when connecting these points, so one should be wary. If you are in doubt as to the correctness of your graph, try plotting a few additional points. Incidentally, note that different scales were used on the axes in Figure 8.3. If we had used the same scale on both axes, we could have plotted only a very few points on the page. After the table of values has been constructed, you should have a better idea as to the appropriate scales to use in order to get a more complete picture on the paper.

Example 2. Graph the function

$$f(x) = \frac{2}{x - 1}.$$

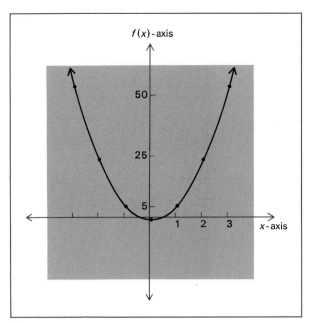

Figure 8.3

Solution: We begin by making a table of some values of $f(x)$ corresponding to convenient values of x:

x	-5	-3	-1	0	$\frac{1}{2}$	$\frac{2}{3}$	1	$\frac{3}{2}$	2	3	4	5
$f(x)$	$\frac{-1}{3}$	$\frac{-1}{2}$	-1	-2	-4	-6	No value	4	2	1	$\frac{2}{3}$	$\frac{1}{2}$

It is very important to note that there is no point of the graph that has an x-coordinate equal to 1. This means that the graph cannot cut the vertical line $x = 1$, so it must consist of two separate branches. We have plotted these points and connected them with the correct curves in Figure 8.4. We call the line $x = 1$ a **vertical asymptote** for the curve, and we call the x-axis a **horizontal asymptote** for the curve. Each branch of the graph gets closer and closer to these lines, but neither branch ever intersects either of the asymptotes.

Plotting points to find the graph of a function is often the easiest way to find the graph, but the method is full of pitfalls. The trouble is that between some pair of plotted points, the graph may behave in some way that is not made apparent by the points actually plotted. One must be sure to plot enough points to identify the graph correctly.

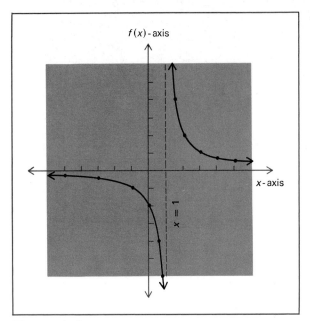

Figure 8.4

Exercises 8.3

*1. Functions of the general form $f(x) = a_n x^n + a_{n-1} x^{n-1} + \cdots + a_1 x + a_0$, where n is a whole number and the coefficients a_n, a_{n-1}, ..., a_1, and a_0 are real numbers, are called **polynomial functions.** The graph of a polynomial function is a one-piece smooth curve. Sketch the graph of the following polynomial functions by plotting points for integer values of x from -5 to 5 and then by plotting more points as needed.
  (a) $f(x) = x^2 - 2x - 5.$ (b) $f(x) = x^3 - 3x^2 - 4x + 10.$
  (c) $f(x) = x^4 - 2x^2.$
  From these three examples can you spot any connection between the highest power of x occurring in the expression for the function and the way the graph moves up and down?

*2. Functions that have the form $f(x) = a^x$, where a is a real number, are called **exponential functions** and their graphs are one-piece smooth curves. Graph the following exponential functions on the same coordinate plane. Use a small scale on the $f(x)$-axis. Plot values of x from -6 to 6 and use the fact that by definition $a^{-6} = 1/a^6$, $a^{-5} = 1/a^5$, ..., $a^{-1} = 1/a^1.$
  (a) $f(x) = 1^x.$ (b) $f(x) = 2^x.$
  (c) $f(x) = 3^x.$ (d) $f(x) = 4^x.$

3. Does the graph of the function $f(x) = 1/(x - 1)$ look like the curve on the left or the curve on the right?

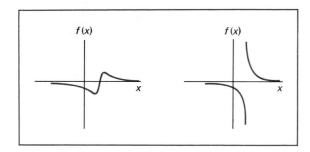

4. Each of the following functions has a graph consisting of two smooth branches. To find these graphs plot points between -4 and 4 as needed. Each of these graphs has both vertical and horizontal asymptotes.
(a) $f(x) = 1/x$.   (b) $f(x) = x/(x - 1)$.

5. Sketch the graphs of the following functions between $x = -10$ and $x = 10$. Plot points as needed.
(a) $f(x) = [\![x]\!]$.   (b) $f(x) = |x|$.

8.4 Applications of Graphing

The graph of a function may be used to reveal information about the function itself. In this section we shall use the graphs of functions in the solution of certain problems involving functions.

Our first application involves using the graph of a function to find the point at which the function is minimal.

Example 1.   It is desired to construct a rectangle encompassing an area of 4 square inches. What should be the dimensions of the rectangle in order that it will have the smallest possible perimeter?

Solution: Let the length of the rectangle be represented by x (see Figure 8.5). Then, since the area encompassed by the rectangle is 4, the width of the rectangle

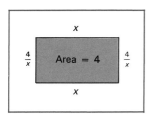

Figure 8.5

will be $4/x$. The perimeter of a rectangle of dimensions x by $4/x$ is $x + x + 4/x + 4/x$ or $2x + 8/x$. That is, we are able to express the perimeter of the rectangle as a function of the length of the rectangle. Representing this function by P, we can express this functional relationship by the equation

$$P(x) = 2x + \frac{8}{x}.$$

Our problem now is to find that value of x for which $P(x)$ is as small as possible. We can find this value of x, approximately at least, by graphing this function and locating its minimum by inspection. The graph of this function is shown in Figure 8.6. Notice that we have plotted points only to the right of the $P(x)$-axis, because negative values of x are not meaningful in this problem (since lengths cannot be negative), so we need not concern ourselves with the part of the graph to the left of the $P(x)$-axis. Happily, we plotted the point for which $x = 2$, and this point appears to be a minimum point. That is, the point $(2, 8)$ appears from our sketch to be the minimum point of this graph. To test this, we might want to try a couple of values of x very near to 2 and on either side of 2 to make sure that their functional values are greater than 8. We compute and find that $P(1.9) = 8\frac{1}{95}$, which is greater than 8, and that $P(2.1) = 8\frac{1}{105}$, which is also greater than 8. Therefore, we may feel rather secure in our conclusion that $(2, 8)$ is the minimum point. This means that the smallest perimeter that a rectangle of area 4 square inches can

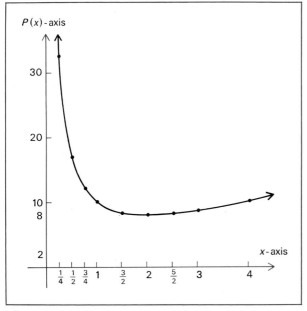

x	$P(x)$
$\frac{1}{4}$	$32\frac{1}{2}$
$\frac{1}{2}$	17
$\frac{3}{4}$	$12\frac{1}{6}$
1	10
$\frac{3}{2}$	$8\frac{1}{3}$
2	8
$\frac{5}{2}$	$8\frac{1}{5}$
3	$8\frac{2}{3}$
4	10

Figure 8.6

have is 8 inches. In this case the width of the rectangle is $\frac{4}{2}$ or 2 inches. This means that the rectangle that we should use is a square.

Example 1 illustrates the general fact that to encompass a given area with a rectangle of the least perimeter, use a square. The next example involves finding a maximum point.

Example 2.   We have a piece of tin measuring 20 inches by 20 inches. It is required to make this into an open-topped box by cutting squares out of the corners and bending up the sides (see Figure 8.7). To the nearest half inch, what should be the dimensions of the pieces to be cut out in order that the resulting box will have the greatest possible volume?

Solution: If we represent the side of one of these smaller squares by x, then the volume of the resulting box is given by the expression

$$(20 - 2x)(20 - 2x)(x).$$

Thus the volume of this box can be expressed in terms of the length x and is therefore a function of x. If we represent this function by the letter V, then it can be described by the equation

$$V(x) = (20 - 2x)(20 - 2x)(x).$$

Our job now is to find that value of x for which $V(x)$ is as large as possible. We can get an approximate value for x by graphing the function and observing its maximum point. This function is graphed in Figure 8.8. Because we are concerned only with values of x between 0 and 10 (no other values make sense in this problem), we need bother only with that portion of the graph between $x = 0$ and

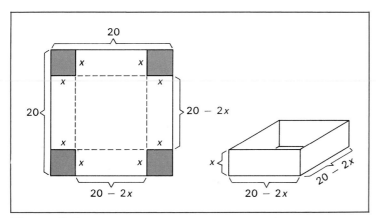

Figure 8.7

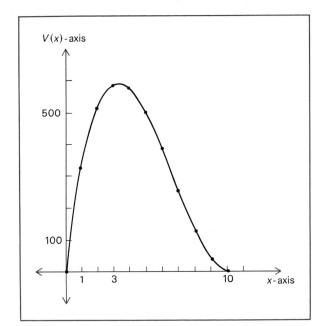

x	V(x)
0	0
1	324
2	512
3	588
4	576
5	500
6	384
7	252
8	128
9	36
10	0

Figure 8.8

$x = 10$. On this portion of the graph, there appears to be a maximum in the vicinity of $x = 3$. Since we want to find x to the nearest half inch so that $V(x)$ is maximum, let us compute $V(2.5)$ and $V(3.5)$:

$$V(2.5) = (20 - 5)(20 - 5)(2.5) = 562.5$$

and

$$V(3.5) = (20 - 7)(20 - 7)(3.5) = 591.5.$$

Since $V(3) = 588$, we see that, to the nearest half inch, $V(x)$ is greatest when $x = 3.5$ inches. This means that if the four corner squares are of dimensions 3.5 inches by 3.5 inches, then the resulting box will have the largest volume—591.5 cubic inches. [In order to be doubly sure that we have the correct answer, you might want to determine $V(4)$ also.]

One of the important uses of calculus is to solve such problems as these and obtain exact answers. Using elementary calculus, we could easily and quickly determine that our answer in Example 1 is exact and our answer to Example 2 is not. The exact dimensions of the squares to be removed in Example 2 should be $3\frac{1}{3}$ inches instead of $3\frac{1}{2}$ inches; however, we asked for the dimensions to the nearest half inch so that our answer is correct subject to the requirements of the example.

Exercises 8.4

1. What is the largest area that can be enclosed by a rectangle of perimeter 100 inches? [*Hint:* Express the area of the rectangle, $A(x)$, in terms of the length x of the rectangle. Plot the values $x = 0, 5, 10, \ldots$, and then plot more values of x as needed.]

2. In order to send a package through the mails, the combined length and girth must not exceed 84 inches. Suppose we want to construct a box with a square cross section that has as large a volume as possible subject to the 84-inch restriction. To the nearest inch what should be the dimensions of the package? (*Hint:* Express the volume as a function of the length of a side of the square end. When you graph the function to find its maximum value, begin by graphing even integer values of x from 0 to 20. Then plot more points as necessary.)

3. We want to make a covered box to hold 64 cubic inches. The box is to have a square base. What should be the dimensions of the box if it is to have a minimal surface area? [*Hint:* Express the surface area in terms of the length of a side of the square bottom. Graph this function $A(x)$ by plotting some points corresponding to integral values of x. Then plot more points as needed.]

4. A book is to be designed, and it has been decided that each page of the book should contain 32 square inches of print. There should be right and left margins of $\frac{1}{2}$ inch, a top margin of $\frac{3}{4}$ inch, and a bottom margin of $1\frac{1}{4}$ inches. What should be the dimensions of a page of this book in order that the least amount of paper be used?

8.5[†] Trigonometric Functions

Trigonometric functions are an important class of functions, and their study is the central part of trigonometry. In this section and the next we shall discuss some of these functions.

The first trigonometric function we shall study is called the **sine function.** This function is defined using a circle and a moving radius. In Figure 8.9 we have shown a circle of radius 1 unit, situated so that its center lies at the origin of the coordinate plane. The segment $\overline{OP}$ is a typical radius of this circle. This radius makes an angle α with the positive x-axis.[1] As the angle α varies, the point P moves along the circumference of the circle. We want to center our attention upon the y-coordinate of the point P. As the angle α varies, say from 0 to 90°, the y-coordinate of the point P varies in turn. In fact, the y-coordinate of the point P is a function

---

[1] By convention, an angle is termed positive if it is measured from the positive x-axis in a counterclockwise direction. Angles measured from the positive x-axis in a clockwise direction are termed negative angles.

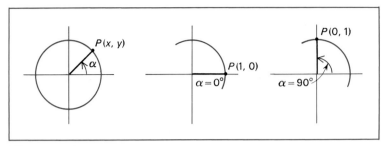

Figure 8.9

of the angle α. For example, when $\alpha = 0°$, the radius $\overline{OP}$ is coincident with the positive x-axis, the coordinate of P is $(1, 0)$, and so the y-coordinate of P is 0. When $\alpha = 90°$, the radius $\overline{OP}$ is coincident with the positive y-axis, the coordinate of P is $(0, 1)$, and so the y-coordinate of P is equal to 1 (see Figure 8.9). As the angle α varies from 0 to $90°$, the y-coordinate of the point P varies from 0 to 1. Figure 8.10 will help you to visualize the way that the y-coordinate varies as α varies from 0 to $90°$. The curve in Figure 8.10 is the graph of the functional relationship between the y-coordinate of P and the angle α for values of α from 0 to $90°$.

The function just described is called the sine function and is symbolized by the special symbol **sin.** For example, since when $\alpha = 0°$, the y-coordinate of P is 0, we can write $\sin 0° = 0$. Also, since when $\alpha = 90°$, the y-coordinate of P is 1, we write $\sin 90° = 1$. In general, we write

$$\sin x = y.$$

That is, the sine of an angle α is the y-coordinate of the point P when the radius $\overline{OP}$ makes an angle α with the positive x-axis.

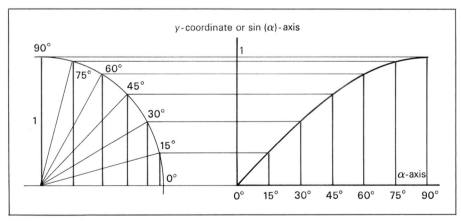

Figure 8.10

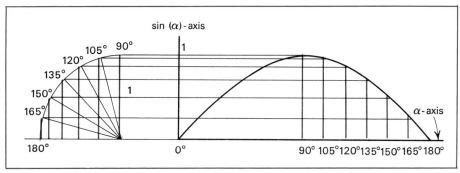

Figure 8.11

Let us go on and see what this function looks like for other values of α. What happens to this y-coordinate as α varies from 90 to 180°? The sine of 180° is 0, since when $\alpha = 180°$ the radius $\overline{OP}$ is coincident with the negative x-axis, the coordinate of P is $(-1, 0)$, and the y-coordinate of P is 0. Hence, as α varies from 90 to 180°, the y-coordinate (that is, sin α) varies from 1 (at 90°) to 0 (at 180°). See Figure 8.11, which shows the graph of the sine function from 0 to 180°. The sine of 270°, sin 270°, is equal to -1, since when $\alpha = 270°$, the radius $\overline{OP}$ is coincident with the negative y-axis and the y-coordinate of P is -1. The sine of 360° is 0 since, when $\alpha = 360°$, the radius $\overline{OP}$ is again coincident with the positive x-axis. Hence, as α varies from 180 through 270 to 360°, the y-coordinate of P varies from 0 through -1 to 0 again. The graph in Figure 8.12 is the graph of the sine function for values of α from 0 to 360°.

Table 8.1 contains numerical values of the sine function for integral values of α from 0 to 90°. With a few exceptions these values have been rounded off to three decimal places.

In discussing the sine function, we centered our attention on the y-coordinate

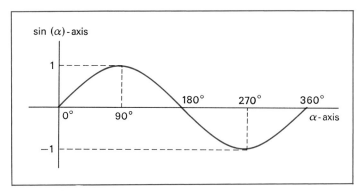

Figure 8.12

Table 8.1

Angle	Sine	Angle	Sine	Angle	Sine	Angle	Sine
0°	.000	23°	.391	46°	.719	69°	.934
1°	.017	24°	.407	47°	.731	70°	.940
2°	.035	25°	.423	48°	.743	71°	.946
3°	.052	26°	.438	49°	.755	72°	.951
4°	.070	27°	.454	50°	.766	73°	.956
5°	.087	28°	.469	51°	.777	74°	.961
6°	.105	29°	.485	52°	.788	75°	.966
7°	.122	30°	.500	53°	.799	76°	.970
8°	.139	31°	.515	54°	.809	77°	.974
9°	.156	32°	.530	55°	.819	78°	.978
10°	.174	33°	.545	56°	.829	79°	.982
11°	.191	34°	.559	57°	.839	80°	.985
12°	.208	35°	.574	58°	.848	81°	.988
13°	.225	36°	.588	59°	.857	82°	.990
14°	.242	37°	.602	60°	.866	83°	.993
15°	.259	38°	.616	61°	.875	84°	.995
16°	.276	39°	.629	62°	.883	85°	.996
17°	.292	40°	.643	63°	.891	86°	.998
18°	.309	41°	.656	64°	.899	87°	.9986
19°	.326	42°	.669	65°	.906	88°	.9994
20°	.342	43°	.682	66°	.914	89°	.9998
21°	.358	44°	.695	67°	.920	90°	1.000
22°	.375	45°	.707	68°	.927		

of the point P. If we had instead centered attention on the x-coordinate, another function would have been defined, called the **cosine function**. Given an angle α, then the cosine of α, cos α, is defined to be the x-coordinate of the point P when the radius $\overline{OP}$ is making an angle of α with the positive x-axis. See Figure 8.13.

Example 1.   What is the cosine of 0°?

Solution: We see from Figure 8.14 that when $\alpha = 0°$, the x-coordinate of the point P is 1. Hence the cosine of 0° is 1, and we write cos 0° = 1.

Example 2.   Find the cosine of 90°, 180°, 270°, and 360°.

Solution: We can use Figure 8.14 to see that the x-coordinate of P, when $\alpha = 90°$, is equal to 0. Hence cos 90° = 0. Cos 180° = -1, since the x-coordinate of P, when $\alpha = 180°$, is -1. Cos 270° = 0, and cos 360° = 1.

The graph of the cosine function for values of α from 0 to 360° is shown in Figure 8.15.

Table 8.2 lists approximate values of the cosine function for values of α from 0 to 90°. With three exceptions, these values have been rounded off to three decimal places.

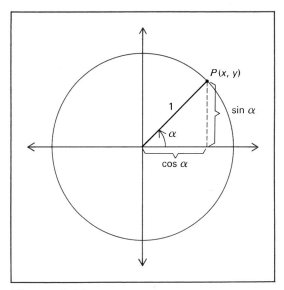

Figure 8.13

Beside the sine and cosine there are four other trigonometric functions called the **tangent, cotangent, secant,** and **cosecant** functions. All of these functions are defined in terms of a circle, as were the sine and cosine functions; for this reason these functions are sometimes called "circular functions." We shall not study these other functions.

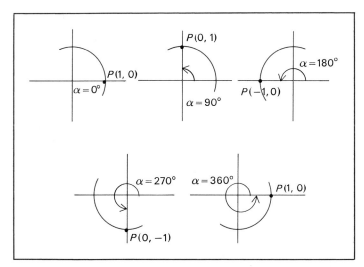

Figure 8.14

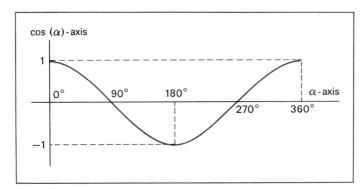

Figure 8.15

Table 8.2 Table of Cosines

Angle	Cosine	Angle	Cosine	Angle	Cosine	Angle	Cosine
0°	1.000	23°	.920	46°	.695	69°	.358
1°	.9998	24°	.914	47°	.682	70°	.342
2°	.9994	25°	.906	48°	.669	71°	.326
3°	.9986	26°	.899	49°	.656	72°	.309
4°	.998	27°	.891	50°	.643	73°	.292
5°	.996	28°	.883	51°	.629	74°	.276
6°	.995	29°	.875	52°	.616	75°	.259
7°	.993	30°	.866	53°	.602	76°	.242
8°	.990	31°	.857	54°	.588	77°	.225
9°	.988	32°	.848	55°	.574	78°	.208
10°	.985	33°	.839	56°	.559	79°	.191
11°	.982	34°	.829	57°	.545	80°	.174
12°	.978	35°	.819	58°	.530	81°	.156
13°	.974	36°	.809	59°	.515	82°	.139
14°	.970	37°	.799	60°	.500	83°	.122
15°	.966	38°	.788	61°	.485	84°	.105
16°	.961	39°	.777	62°	.469	85°	.087
17°	.956	40°	.766	63°	.454	86°	.070
18°	.951	41°	.755	64°	.438	87°	.052
19°	.946	42°	.743	65°	.423	88°	.035
20°	.940	43°	.731	66°	.407	89°	.017
21°	.934	44°	.719	67°	.391	90°	.000
22°	.927	45°	.707	68°	.375		

Exercises 8.5

1. By examining the graphs of the sine and cosine functions between 0 and 90°, you can see that these graphs intersect. Using the table of values, find the angle at which they intersect; that is, find α such that $\sin \alpha = \cos \alpha$.

2. In the table of values of the sine function we rounded all values off to three decimal places to the right of the decimal point, except for the values of the angles 87°, 88°, and 89°, which were rounded off to four decimal places. Why were we forced to round these values off to four places rather than to three places as we did for all the other values? (This same situation arises in the table of values for the cosine function.)

3. By analyzing the way the y-coordinate of the point P varies as α varies from 0 to 360°, we obtain the graph shown in Figure 8.12 for the sine function. How does this graph look for values of α from 360 to 720°?

4. In Figure 8.15 we have sketched the graph of the cosine function for values of α from 0 to 360°. What does this graph look like for values of α from 360 to 720°?

5. Prove that for any angle α between 0 and 90° the equation

$$(\sin \alpha)^2 + (\cos \alpha)^2 = 1$$

is true. (Actually, this is a theorem and is true for all values of the angle α. Study Figure 8.13 to find the proof.)

8.6† Applications of the Trigonometric Functions

In this section we shall use the sine and cosine functions to solve certain kinds of problems involving the determination of distances that would be difficult or even impossible to find directly.

Consider the right triangle shown in Figure 8.16. This triangle has a hypotenuse of length 1 unit. Using the sine and cosine functions, we can express the lengths of the two legs of this triangle in terms of the angle α. To see how to do this, place the right triangle inside a circle, as shown in Figure 8.17. This circle has radius 1 unit and so it follows that the vertical leg of the right triangle has length $\sin \alpha$ (since the length of this vertical leg is the y-coordinate of the point P). Also,

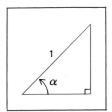

Figure 8.16

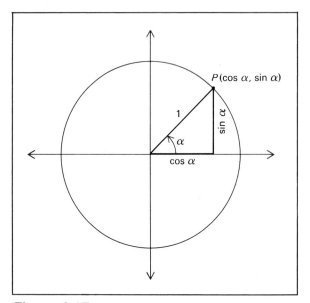

Figure 8.17

since the length of the horizontal leg is the x-coordinate of the point P, the length of this leg is equal to cos α.

Now consider a right triangle whose hypotenuse has length k units (k a real number). Again the lengths of the legs of this triangle can be expressed in terms of the angle α. We use the fact that this triangle (Figure 8.18) is similar to the triangle in Figure 8.17, and so the sides of the two triangles have proportional lengths. Thus, since the hypotenuse of one triangle has length that is k times the length of the other triangle, the lengths of the legs of the one triangle are k times the lengths of the other triangle. This means that the vertical leg in Figure 8.18 has length k (sin α) and the horizontal leg has length $k(\cos \alpha)$. In the following examples we shall use this information to find inaccessible distances.

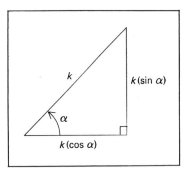

Figure 8.18

Example 1. A man lives in a cabin located on the side of a mountain at the base of which is a lake. It is known that the elevation of the lake is 6000 feet. The man must walk up the mountain a distance of 1 mile to reach his cabin from the lake. Also, the angle of incline of the mountain from the lake to the cabin is about 18°. What is the approximate elevation of the cabin?

Solution: In Figure 8.19 we have drawn a sketch that conveys the information supplied in this problem. Also, we have drawn the right triangle that is the mathematical model of the problem. If we call the vertical distance of the cabin above the lake level x, then, because this distance x is the length of the leg opposite the 18° angle, we can express x as a function of α by the equation

$$x = (5280)(\sin 18°).$$

Using Table 8.1, we observe that $\sin 18° = .309$ so that

$$x = (5280)(.309)$$
$$= 1631, \text{ approximately.}$$

Hence the elevation of the cabin is $6000 + 1631$ or 7631 feet, approximately.

Example 2. Two towers are located on either side of a river (see Figure 8.20). From a point 2000 yards downstream of one of the towers, the tower on the opposite side of the river makes an angle of 22°. How far apart are the two towers?

Solution: The sketch in Figure 8.20 contains all of the given information for this problem and a drawing of the right triangle that is the mathematical model of the problem. Let the distance between the towers be called x. If we knew the length of the hypotenuse of this triangle, then we could determine the length x by using the equation

$$x = (\text{length of hypotenuse})(\sin 22°).$$

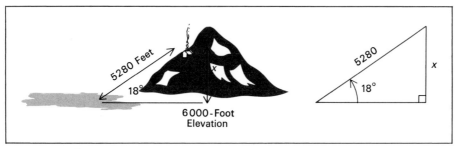

Figure 8.19

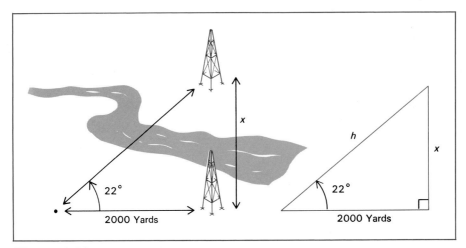

Figure 8.20

But we were not given the length of the hypotenuse. Call this length h. We can determine h by using the other information given by the problem. That is, we know that the length of the leg adjacent to the 22° angle is equal to the length of the hypotenuse times the cosine of 22°. Hence

$$2000 = (h)(\cos 22°).$$

From Table 8.2, $\cos 22° = .927$, so

$$2000 = (h)(.927)$$

or

$$h = \frac{2000}{.927} = 2161 \text{ yards, approximately.}$$

Now that we know h, we can find x:

$$x = (2161)(\sin 22°)$$
$$= (2161)(.375)$$
$$= 810 \text{ yards, approximately.}$$

We have found the distance between the towers indirectly. It would be difficult to find this distance without using trigonometry.

Exercises 8.6

1. A wire has been stretched from the top of a 1676 foot tall TV tower to the ground. The length of wire required was 2000 feet. The installer was supposed to measure the angle the wire makes with the top of the tower but he forgot. Can he determine this angle without climbing to the top of the tower?

2. A tree casts a shadow 92 feet long at a time of day when the elevation of the sun is 48°. How tall is the tree? (*Hint:* You will first have to determine the length of the line from the top of the tree to the end of the shadow.)

3. By making observations from earth it can be determined that the angle α shown in the drawing below is equal to about 1°. If you assume the radius of the earth is approximately 3963 miles, what is the approximate distance between the center of the earth and the center of the moon? Actually, the true angle subtended is a bit less than 1°; therefore, is the answer you got too big or too small?

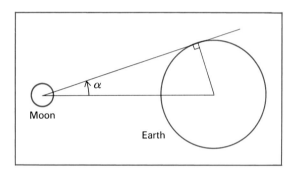

4. A 300-foot radio tower has been installed. It is necessary to run a supporting cable from the top of the tower to a point on the ground 80 feet away from the base of the tower. In order to install the ground anchor for this cable, it is necessary to know the angle the cable will make with the ground. What is this angle? (*Hint:* First compute the length of cable using the Pythagorean Theorem. Then use either the sine or cosine function and the appropriate table of values to find the approximate angle.)

5. At exactly 10 A.M. on a certain day a tree 60 feet tall casts a shadow 24 feet long. What was the angle of elevation of the sun at that time?

6. The leaning tower of Pisa is about 179 feet high and is about 16.5 feet out of plumb. Find the angle the tower is off from vertical.

The Theory of Probability

It is always difficult to establish just when a particular part of mathematics got its start, but the theory of probability is supposed to have started in 1654. In this year a gambler, Chevalier de Mere, asked the mathematician Blaise Pascal (French, 1623–1662) about a certain dice game that was giving him trouble. Pascal corresponded with Fermat, and the theory of probability stems from their correspondence concerning this dice game and related ideas. Today the theory is an important part of mathematics and has significant applications to all fields of science.

This story about the origin of probability is pleasant enough, and it may even be true. At any rate, it is frequently found in the literature. It is true that probability spent its early years in gambling casinos. In fact, it still spends a part of its time there, in the sense that gambling problems provide excellent illustrative examples to use in discussing elementary probability. We shall employ such games of chance throughout our discussion of probability.

9.1 A Priori and Statistical Probability

There are two kinds of probability, and you have had informal experience with both of them. They are called "a priori probability" and "statistical (or a posteriori)

245

probability." In this section we shall differentiate between the two and begin a study of a priori probability.

The difference between a priori and statistical probability can perhaps best be illustrated by looking at some examples. Consider these two problems, each of which has to do in some way with probability.

Problem 1. A coin is tossed and caught. What is the probability that the coin will come up heads?

Problem 2. What is the probability that a person born in 1968 in Toledo will live to be 80 years old?

You will surely agree that each of these problems has to do with probability, but that these are two very different kinds of probability problems. Certainly, in order to answer Problem 2, we would have to collect evidence of some kind—we would have to examine birth and death records or evidence of this sort. This problem cannot be answered in a purely mathematical way. But Problem 1 does admit a purely mathematical solution: There are exactly two ways the coin can come up when it is tossed. Each of these two ways is equally likely. Exactly one of these two ways will result in a head. Therefore, the chances of tossing a head are 1 in 2. The mathematician expresses this by saying that the a priori probability of tossing a head is $\frac{1}{2}$. The prefix words **a priori** mean "valid independently of observation," and this is the way we solved Problem 1. We did not need to find a coin and actually toss it many times in order to determine the probability. But in Problem 2, no such purely mathematical solution is possible. Problem 2 can be solved only through observation. The kind of probability we are talking about in Problem 2 is **a posteriori** ("based upon actual observation or upon experimental data") or **statistical** probability.

We should remark that under certain conditions we might feel it important to regard Problem 1 as a problem in statistical probability. For example, if we felt that the coin being tossed might be biased, then we would want to experiment with the coin in order to determine the probability. If we tossed the coin, say, 1000 times and found that it came up heads only 100 times, then we would be strongly tempted to disregard the a priori probability as being not applicable to this biased coin and use the statistical probability of $\frac{100}{1000}$ instead. Generally, however, we regard coins and the like as being unbiased and apply a priori probability.

To introduce the various terms that we shall be working with in our study of a priori probability, let us consider the throwing of a single die. (A **die** is a perfect cube the faces of which are inscribed with the digits 1 through 6.) When the die is thrown, there are six possible **outcomes:** that a 1 will come up, that a 2 will come up, and so on. These possible outcomes are called **mutually exclusive** because it is impossible for more than one of them to occur at the same time. Also, under the assumption that the die is balanced, these six different outcomes are **equally likely.** This means that the die has been constructed in such a way that no one number will come up more often than any other number over a long series of throws. If the die were loaded, then the possible outcomes would not be equally likely, although they would still be mutually exclusive. We shall always assume in our work with a priori probability that the die is perfect. We refer to the throwing of a die as an **experiment.**

Another example of an experiment, the outcomes of which are mutually exclusive and equally likely, is the tossing of a coin. Still another example is the dealing of a card from a well-shuffled deck of cards.

Suppose now that we are going to perform some experiment whose outcomes are mutually exclusive and equally likely and that we have already identified each of the (finitely many) possible outcomes. Then the **a priori probability** that a given event will occur as the result of the experiment is given by the quotient f/n, where n represents the total number of possible outcomes of the experiment and f represents the total number of these that are favorable to the event's occurring. Here are some examples that will help to clarify this definition.

Example 1. What is the probability that, in throwing a die, an even number will come up?

Solution: The total number of possible outcomes is six. The event we are talking about is that an even number will come up and there are three outcomes that are favorable to this event's occurring. These are the outcomes that a 2 will come up, that a 4 will come up, and that a 6 will come up (see Figure 9.1). All the other outcomes are unfavorable. Hence the probability is $\frac{3}{6}$ or $\frac{1}{2}$.

Example 2. An American roulette wheel contains the numbers 1 through 36 (evenly divided into red and black numbers—see Figure 9.2) and two green

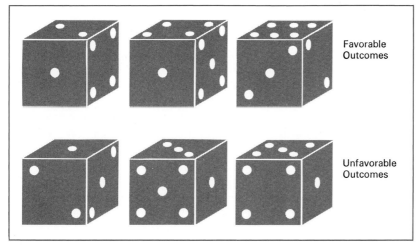

Favorable Outcomes

Unfavorable Outcomes

Figure 9.1

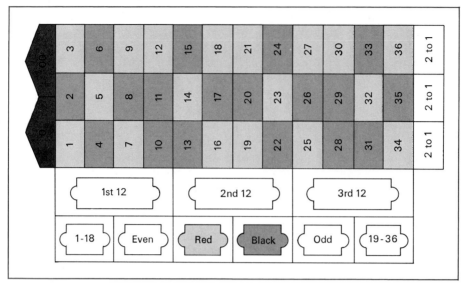

Figure 9.2

numbers 0 and 00 (dark red here).[1] What is the probability that when the wheel is spun, the little white ball will fall into 0 or 00?

Solution: There are a total of $36 + 2 = 38$ outcomes that may result from the experiment of spinning the wheel. Of these 38 outcomes, exactly two are favorable to the outcomes 0 or 00 being selected. Hence the probability of getting a 0 or 00 is $\frac{2}{38} = \frac{1}{19}$.

Example 3. What is the probability of drawing the 5♠ from a deck if the A♠, 2♠, 3♠, and 4♠ have already been removed?

Solution: The experiment consists of dealing a card from the 48-card deck that remains after the four specified cards have been removed. Of the 48 possible outcomes of such a deal only one, the 5♠, will be favorable. Hence the probability is $\frac{1}{48}$.

Example 4. What is the probability of completing the hand 2♠, 3♠, 4♠, and 5♠ to make a flush? (*Note:* A "flush" in poker is a hand consisting of five cards all of the same suit.)

Solution: We have depleted the deck by four spades, so there are only nine spades left in the 48-card depleted deck. Hence the probability of drawing a fifth

---

[1]The American wheel differs from the European wheel in that the European wheel contains only one green number, 0.

spade is $\frac{9}{48}$ or $\frac{3}{16}$. That is, there are 48 ways of dealing off the fifth card, but only nine of them are favorable to dealing off a spade.

The extreme limits of probabilities are 0 (impossibility) and 1 (certainty). All other probabilities lie somewhere between these limits. The nearer a probability is to 0 the less likely it is that the event will occur, and the nearer the probability is to 1 the more likely it is that the event will occur.

The examples of this section were all rather simple. In order to work more complicated probability problems, we shall have to develop some counting techniques. We shall do this in Section 9.3. First, however, we pause to discuss the connection between a priori probability and the real world.

Exercises 9.1

1. Compute these probabilities having to do with the rolling of two dice by listing all possible outcomes and listing all favorable outcomes.
   (a) What is the probability that a number less than six will be rolled?
   (b) What is the probability that an even number will be rolled?
   (c) What is the probability that an even number less than six will be rolled?
   (d) What is the probability that either an even number or a number less than three will be rolled?

2. The following questions have to do with the American roulette layout shown in Figure 9.2.
   (a) If you place your bet on the space marked "1–18," then you will win if any one of the numbers 1 through 18 comes up. What is the probability of this event's occurring?
   (b) If you put your bet on the space marked "First 12," you will win if any one of the numbers 1 through 12 comes up. What is the probability of this event's occurring?
   (c) If you place your bet on one of the spaces marked "2 to 1," you will win if any one of the numbers in the column in which you have placed your bet comes up. What is the probability of this event's occurring?
   (d) Compare the probabilities for *red, black, odd,* and *even.* (*Note:* 0 and 00 are red numbers that are neither odd nor even.)
   (e) It is possible to place your bet on the lower left corner of the square containing the number 1. By so doing, you bet that one of the numbers 0, 00, 1, 2, and 3 will come up. What is the probability of winning with this bet?
   (f) It is possible to place your bet on the common corner of the squares containing the numbers 22, 23, 25, and 26. By so doing you bet that one of these four numbers will come up. What is the probability of winning on this bet?

3. A single card is drawn from a deck. What is the probability that it is either a heart or a face card?

4. Would you rather bet the number 10 on an American or on a European roulette wheel? Explain using probability.

5. Describe the different possible outcomes of each of the following experiments. Are these outcomes mutually exclusive? Equally likely?
   (a) Selecting a card from a shuffled deck of cards.
   (b) Choosing a wife.
   (c) Asking a supermarket cashier for change for a quarter.

6. You have been dealt the cards 2♥, K♦, J♠, and 9♣ from a shuffled deck. What is the probability that the next card dealt to you will give you a pair? (A *pair* is two cards of the same rank.)

7. In poker a *straight* consists of five cards whose ranks are in order, for example, 2, 3, 4, 5, and 6. Ace counts high and low. If from a complete deck you have been dealt the cards shown below, what is the probability that upon being dealt a fifth card you will fill to a straight? (We shall ignore suits in this problem.)
   (a) A, 3, 4, 5.   (b) A, 2, 3, 4.   (c) 2, 3, 4, 6.

8. One way to alter the behavior of a die is to shave a small amount of surface material off the faces containing the six and the one. (These are on opposite sides of the die.) Such a die is called a *Six-Ace Flat*. Will the six possible outcomes of throwing a Six-Ace Flat be mutually exclusive? Which outcomes will be more likely and which will be less likely?

9. A couple has two children (not twins). There are four equally likely ways that these two children could have been born: Boy-boy, girl-girl, boy-girl, and girl-boy.
   (a) It is known that one of the children is a boy. What is the probability that the other is also a boy?
   (b) It is known that the youngest of the two children is a boy. What is the probability that the other is also a boy?

10. The cards K♥, Q♥, 6♥, and 7♥ are removed from a deck and this four-card partial deck is well shuffled. Each of two players is dealt two cards. One of the players, Ethelred, says, "I have a face card in my hand." At another table another couple has gone through exactly the same procedure, and one of these players, Ethelbert, says, "I have the king of hearts in my hand." Which of these two players, Ethelbert or Ethelred, is more likely to have two face cards in his hand? Try to predict which has the greater probability before working the problem. (*Hint:* First list all of the possible hands: K♥ and Q♥, K♥ and 6♥, K♥ and 7♥, and so forth.)

9.2 The Law of Large Numbers

The law of large numbers lies at the heart of the theory of probability and is one of the fundamental theorems in this subject. We shall discuss the theorem informally, without giving a precise statement of it.

We know that the a priori probability of tossing a head with a perfect coin is $\frac{1}{2}$. When we say this, we must clearly appreciate that all we mean is that of the two possible outcomes of the toss, one of them is favorable to a head's being tossed. The **law of large numbers** is a theorem that relates this purely mathematical statement of the a priori probability to what might actually happen when a coin is tossed a great many times. The law consists of two parts. The first part states simply that, by tossing the coin sufficiently many times, we can make the ratio of heads to total tosses be as near to the a priori probability of $\frac{1}{2}$ as we desire. That is, if we let N denote the total number of tosses and h_N denote the total number of heads obtained in these N tosses, then by making N large enough we can assure that the ratio h_N/N will be as near to $\frac{1}{2}$ as we desire.

For example, suppose that we want to make the ratio h_N/N be within .001 of $\frac{1}{2}$. We can make this happen if we toss the coin sufficiently many times—if we make N sufficiently large. Note that we have not said how many times we must toss the coin before the ratio of heads to total tosses will be in the range .499 to .501; we have only said that there exists some number of tosses such that, after that many tosses, the ratio will be in the range .499 to .501. It is impossible to predict how many tosses might be required. We could not make a statement such as, "After one trillion tosses the ratio h_N/N will be in the range .499 to .501." The reason we could not make such a statement is obvious: It is conceivable that in the first trillion tosses a tail would be tossed every time. The probability that this would happen is very small indeed, but there is a definite nonzero probability that it could happen.

The first part of the law of large numbers is often called the **law of averages.**

The second part of the law of large numbers deals with the difference between the number of heads actually obtained after N tosses and the number of heads we would be led to anticipate on the basis of the a priori probability. Since the a priori probability of tossing a head is $\frac{1}{2}$, we might expect that after N tosses, we would have obtained $N/2$ heads. But we know that in actual practice it is unlikely that we would obtain $N/2$ heads after N tosses. The actual number of heads could be less than $N/2$ or greater than $N/2$. That is, the difference $N/2 - h_N$ might be positive, actually equal to zero, or negative. The second part of the law of large numbers says that this difference will fluctuate around 0, but that, as the total number of tosses gets larger and larger, this difference will fluctuate more and more wildly between positive and negative values.

The graph in Figure 9.3 shows in a general way how the difference $N/2 - h_N$ jumps up and down as N becomes larger and larger and how these jumps reach lower lows and higher highs. The conclusion we draw from this fluctuation of the

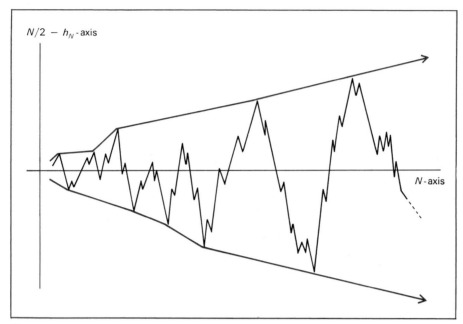

Figure 9.3

difference $N/2 - h_N$ is that as N becomes larger and larger, there will occur longer and longer runs of tails and longer and longer runs of heads. The longer and longer runs of tails are what make the graph attain higher and higher highs as N gets larger. The longer and longer runs of heads are what make the graph attain lower and lower lows as N gets larger and larger.

Thus we see that the law of large numbers (part 2) anticipates long runs of heads and long runs of tails. When such runs occur, we should not really be surprised, because we know that they are bound to occur eventually. For example, betting the red and black on roulette is much the same as betting heads or tails on the flip of a coin. The longest run of one color that we know about was a run of 32 red numbers at Monte Carlo. This event caused considerable commotion at the casino, as indeed it should have. Such long runs happen very rarely. But we should not think that their happening is something truly extraordinary, because the law of large numbers tells us that they are bound to happen. Provided that this world lasts long enough, the law of large numbers tells us that there will sometime be a run of, say, 2000 red numbers at Monte Carlo. The sun may explode and turn Monte Carlo into a cinder 4 billion years before the event is to take place, but if we last long enough it will happen.

The law of large numbers is often misused. To illustrate these misuses, suppose that we were to undertake a series of coin-tossing experiments and that we recorded the results of each toss. Suppose also that on each of the first 32 tosses we got a head. How would you react to this? Very likely you would be surprised that this run of heads happened right at the beginning of the test. After recovering from your shock, you would prepare to make the thirty-third toss. What about this thirty-third toss? Do you have any strong feelings as to what will or ought to happen? There are different points of view here; let us examine each of them.

First, there is the opinion held by a fellow named Skeptic. Skeptic reacts to this series of tosses by stating that he cannot believe that the coin we are using is an unbiased coin. Somehow, Skeptic feels, this coin has been constructed so that it will come up heads every time. Skeptic is reacting to this unusual situation by claiming that, in his opinion, the statistical evidence is such that he for one cannot accept the coin as being perfect. Therefore, he must apply statistical probability rather than a priori probability to this situation, and he concludes that the chances are very high that a head will come up on the thirty-third toss. This is a reasonable attitude to take, but we shall ignore Skeptic because we insist that the coin being used is unbiased—it is either perfect or so nearly perfect that no bias is detectable.

The second attitude is held by Lawyer. Lawyer predicts in no uncertain terms that a tail is much more likely to come up on the thirty-third toss than a head. Lawyer claims that in the long run the number of heads will equal the number of tails and, because heads have come up 32 times so far, according to the law of averages the thirty-third toss will be a tail. Lawyer knows that after sufficiently many tosses of the coin, the ratio of heads to total tosses (h_N/N) will be close to $\frac{1}{2}$. This ratio now stands at $\frac{32}{32}$ or 1. Lawyer feels that the law of averages is going to make this ratio begin its return to $\frac{1}{2}$ *right now*. But this is patent nonsense. The law of averages says no such thing. The law of averages would not be disturbed if heads came up on each of the first million tosses. All the law of averages says is that, after a sufficiently large number of tosses, this ratio will be near $\frac{1}{2}$. How large "sufficiently large" is we do not know and cannot know. In this particular series of tosses, it might be that it will take millions and millions of tosses to get the ratio of heads to total tosses near to $\frac{1}{2}$.

The third attitude is held by Gambler. Gambler knows that the second part of the law of large numbers tells him to expect long runs of heads. He thinks that so unusual a happening as a run of 32 heads must indicate that we are now in the middle of one of these long runs of heads. So he predicts that heads will come up on the thirty-third toss. But Gambler is making the same kind of mistake that Lawyer made. Gambler is trying to draw information about one particular toss of the coin from the law of large numbers, and this simply cannot be done. He and Lawyer are just working opposite sides of the same street.

The last attitude is held by Uninterested. Uninterested is completely bored by the question of what is going to happen on the thirty-third toss, and, while Lawyer and Gambler are arguing and pounding the table to lend credence to their arguments, Uninterested has gone to sleep. He knows that the a priori probability of getting a head on any one toss of the coin is exactly the same as the a priori probability of getting a tail. Hence the coin is just as likely to come up heads as it is to come up tails on the thirty-third toss. Uninterested knows that the coin tossings are independent experiments, in the sense that no one toss is in any way affected or controlled by what happened on earlier tosses. He knows that the coin cannot remember that it has come up heads 32 times already. Perhaps if the coin could remember, it would be properly chagrined and would try very hard to come up tails the thirty-third time, but such speculation is wasted. Uninterested is disgusted by Lawyer's and Gambler's attempts to draw information concerning what is going to happen on one particular toss of the coin from the law of large numbers. He correctly understands that this law will give him information only as to what will happen after a sufficiently large number of tosses, but he also knows that he cannot predict how large "sufficiently large" might be.

Now that we have a good idea of the role played by a priori probabilities, let us go on to consider some counting techniques that we can use to compute probabilities in less trivial situations than coin tossings.

Exercises 9.2

1. We asserted in the text that tossing a coin for heads and tails is very nearly the same as spinning a European roulette wheel for red and black, but these are not really the same experiments. How does the probability of tossing a head differ from the probability of spinning red on a European roulette wheel?

2. Here are the results of 200 coin tossings:
   THHTT HTHTT HTHTH THHTH TTTTH THHHT TTTHH HTHTH
   TTTHT HTHHT THTHH TTHTT HTHTH TTTTH HHHTH TTHHH
   HHHTH HTHTH THHHH THTTT THHTH HTTHH HTHTH THTHH
   HTHTT HHHTH HHTHT HHHTH HHHTT TTTHH HTTHH HHHHT
   HHHTH TTTHH HHTHH TTHTT THHHT TTTHH HHHHT HHTTH
   Starting with $N = 5$, compute the number $N/2 - h_N$ for each value of $N = 5$, 10, 15, 20, 25, ..., 200, and plot the corresponding points $(N, N/2 - h_N)$ as in Figure 9.3. To what extent does this graph resemble the graph shown in Figure 9.3? Are heads occurring in the way that you would expect them to occur?

3. Repeat Exercise 2 by making 200 coin tossings yourself. How do your results compare with those of Exercise 2? (You can save time by tossing a number of coins at once.)

9.3 Some Counting Problems

In this section we shall discuss some counting techniques that can be used to simplify the computation of probabilities. These methods will be used in the next section to compute probabilities in less trivial situations than we encountered in Section 9.1.

All counting techniques are based upon the following **counting principle:** Suppose that two tasks are to be performed. If the first task can be done in n different ways and the second task can then be performed in m different ways, then the two tasks can be performed (in order) in nm different ways. This counting principle can be extended in the obvious way to apply to any finite number of tasks to be performed in order.

Example 1.   How many license plates can be made of the following form?

letter-letter-letter-digit-digit.

Solution: There are 26 ways to fill in the first letter; then there are 26 ways to fill in the second letter; then there are 26 ways to fill in the third letter; then there are 10 ways to fill in the first digit; and, finally, there are 10 ways to fill in the second digit. Hence there are $26 \cdot 26 \cdot 26 \cdot 10 \cdot 10 = 14{,}576{,}000$ different license plates that can be made of this form.

Example 2.   How many license plates can be made of the form

letter-letter-letter-digit-digit

if no letter or digit is to be repeated?

Solution: There are 26 ways to fill in the first letter. But after this letter has been chosen, there remain only 25 letters to use in filling in the second letter. Then there are only 24 ways to fill in the third letter. The first digit can be filled in in any one of 10 ways. But then the second digit can only be filled in in only 9 different ways. Hence there are only $26 \cdot 25 \cdot 24 \cdot 10 \cdot 9 = 1{,}904{,}000$ license plates in which no letter or digit is repeated.

Note the difference between Examples 1 and 2. In Example 1 each task to be performed (each filling in of a space) was **independent** of the tasks that were done previously. But in Example 2 the second task was **dependent** upon what happened as a result of the first task. That is, what could happen as a result of the second task was affected by what already had happened as a result of the first task. Then what could happen as a result of the third task depended upon what already had happened as a result of each of the first and second tasks. As we see illustrated in Example 1, if the tasks to be performed are independent, then the number of ways a series of such tasks can be performed is simply the product of the ways the individual tasks can be performed. But, if the tasks are dependent, then the number of ways the series of tasks can be performed is the product of the numbers of ways each individual task can be performed, assuming that the preceding tasks have already been performed.

Example 3.   The Morse Code is a device by means of which the 26 letters and 10 digits can by symbolized by dots or dashes. Is it possible to devise a code similar to the Morse Code that does not involve any blocks of five or more dots or dashes?

Solution: No. There are two blocks containing exactly one dot or dash. There are $2 \cdot 2 = 4$ blocks containing two dots or dashes. There are $2 \cdot 2 \cdot 2 = 8$ blocks containing exactly three dots or dashes. There are $2 \cdot 2 \cdot 2 \cdot 2 = 16$ blocks containing exactly four dots or dashes. Hence there are $2 + 4 + 8 + 16 = 30$ blocks containing not more than four dots or dashes. Because there are 36 characters to be represented, there must be six blocks containing at least five dots or dashes.

Many counting problems may be reduced to the following "common denominator" kind of problem. Given a finite set containing $n(n > 0)$ objects, how many subsets does it possess that contain $t(0 < t \le n)$ objects? If we can answer this problem, we can solve a wide variety of related counting problems.

Example 4. The set $\{a, b, c\}$ has three two-object subsets: $\{a, b\}$, $\{a, c\}$, and $\{b, c\}$.

Example 5. How many three-object subsets does the set $\{a, b, c, d, e\}$ possess?

Solution: A complete list of the three-object subsets is

$\{a, b, c\}$	$\{a, c, d\}$	$\{a, d, e\}$	$\{b, c, d\}$	$\{b, d, e\}$	$\{c, d, e\}.$
$\{a, b, d\}$	$\{a, c, e\}$		$\{b, c, e\}$		
$\{a, b, e\}$					

These are all of the three object subsets, and so $\{a, b, c, d, e\}$ contains ten three-object subsets.

Solving such problems by listing the subsets is tedious and would be difficult if n were very large. There is a formula that conveniently answers this general problem, a formula that expresses the number of t-object subsets. This formula is most often expressed in terms of the **factorial notation.**[2] If n represents a nonzero whole number, then the symbol $n!$ (read "n factorial") is used to abbreviate the product of n and all the nonzero whole numbers less than n. Thus $1! = 1$, $2! = 2 \cdot 1 = 2$, $3! = 3 \cdot 2 \cdot 1 = 6$, $4! = 4 \cdot 3 \cdot 2 \cdot 1 = 24$, $5! = 5 \cdot 4 \cdot 3 \cdot 2 \cdot 1 = 120, \ldots$, $10! = 10 \cdot 9 \cdot 8 \cdot 7 \cdot 6 \cdot 5 \cdot 4 \cdot 3 \cdot 2 \cdot 1 = 3{,}628{,}880$, and so on. The symbol $0!$ is used as another name for 1. We shall point out the reason for this in a moment.

Using this factorial notation, we find that the number of t-object subsets of a set containing $n(n > 0)$ objects is given by the expression

$$\frac{n!}{t!(n-t)!}.$$

[2]This notation was invented in 1808. Its only purpose is to make typesetting easier.

We shall represent this number by $C(n, t)$, and so we can write the formula

$$C(n, t) = \frac{n!}{t!(n - t)!}.$$

For example, the number of two-object subsets of a set containing six objects is

$$C(6, 2) = \frac{6!}{2!4!} = \frac{6 \cdot 5 \cdot 4 \cdot 3 \cdot 2 \cdot 1}{2 \cdot 1 \cdot 4 \cdot 3 \cdot 2 \cdot 1} = 15.$$

The reason we defined 0! to be another name for 1 was so that this formula would work when $n = t$. When $n = t$, we are asking for the number of n-object subsets of a set containing n objects, and there is only one of these (namely, the set itself). So we see that $C(n, n) = 1$. But according to the formula,

$$C(n, n) = \frac{n!}{n!0!} = \frac{1}{0!}.$$

Thus, in order for the formula to be correct when $n = t$, we must define the symbol 0! to be equal to 1.

Example 6. How many different five-card poker hands can be dealt from a standard deck?

Solution: Reduced to its essentials, all we are asked here is, how many different five-object subsets does a set that contains 52 objects possess? Hence we are required to compute $C(52, 5)$.

$$C(52, 5) = \frac{52!}{5!47!} = \frac{52 \cdot 51 \cdot 50 \cdot 49 \cdot 48}{5 \cdot 4 \cdot 3 \cdot 2 \cdot 1} = 2,598,960.$$

Example 7. A class contains 27 students, and the teacher wants to select 5 students to go to the board and work problems. In how many ways can he select these 5 students?

Solution: We must compute $C(27, 5)$.

$$C(27, 5) = \frac{27!}{5!22!} = \frac{27 \cdot 26 \cdot 25 \cdot 24 \cdot 23}{5 \cdot 4 \cdot 3 \cdot 2} = 80,730.$$

Example 8. If you were a student in the class mentioned in Example 7, how many of the 80,730 five-student sets would contain you?

Solution: We are asked for the number of five-student sets that can be constructed containing you. Starting with you as an object, such a set is formed by including 4 more students in the set. There are 26 students from which to draw these 4 students, and so there are $C(26, 4)$ different ways to complete a five-object set that already has you as its first object. There are 14,950 sets of 5 students that contain you.

The counting techniques discussed here will enable us to solve a surprisingly large number of probability problems, but there are many counting techniques beside these. The study of such techniques is called **combinatorial analysis.** Combinatorial analysis plays a large part in modern mathematics and is particularly important in probability and statistics. Loosely speaking, we may think of combinatorial analysis as the art of counting.

Exercises 9.3

1. A bridge hand consists of 13 cards from a standard deck. How many different bridge hands are possible?
2. You are to select a total of nine numbers from among the numbers 1 through 80 by selecting three numbers from among 1 through 37, three more from among 38 through 59, and three more from among 60 through 80. In how many different ways can you do this?
3. If we place no restrictions upon the internal characteristics of a "word," how many three-letter words are possible in the English language? If we assume that each three-letter word must contain at least one vowel, how many such words can there be? If no three-letter word contains fewer than one or more than two vowels, how many such words can there be?
4. How many divisors does the whole number $2^a \cdot 3^b \cdot 5^c \cdot 7^d \cdot 11^e$ possess? How many divisors does 3300 possess? List all of these divisors.
5. Let us assume that every nation has a flag consisting of one, two, or three of the colors red, white, yellow, blue, green, black, and gold, but that no two nations employ the same combination of colors in their flag. How many nations may be represented by flags of one color? Of two colors? Of three colors? Of three or fewer colors?
6. A man witnessed a hit-and-run accident. Although he was not able to identify the complete license number of the car, he did recognize that the license was of the form

    letter-letter-letter-digit-digit-digit,

    that the license contained the letter K in some position, and that the license involved the combination of digits 17 next to each other in some position. If the police had to check all cars with licenses fitting this description, how many cars would they have to check?

7. Use the counting principle and the notion of prime factorization to prove the following two theorems.
   (a) If a whole number is a perfect square, then it has an odd number of divisors.
   (b) If a whole number is not a perfect square, then it has an even number of divisors.
8. Six lines are drawn in the plane so that no three of them intersect at the same point and no two of them are parallel. How many triangles are contained in this configuration of lines? (*Hint:* Each set of three lines forms a triangle.) How many triangles are formed if exactly two of the lines are parallel?

9.4 Computing Probabilities

We shall continue our study of probability with the study of certain gambling games. It is from such games that the theory derived historically; they still serve as efficient tools with which to learn probability.

We have already discussed the relatively simple game of roulette. Let us begin here with the dice game called craps. The rules for craps are quite simple: One begins by rolling two dice. If either a 7 or an 11 is obtained in the first roll, the shooter wins. If one of the numbers 2, 3, or 12 is obtained, the shooter loses. If some other number, say a 5, is obtained on the first roll, then the number becomes the shooter's "point." On subsequent rolls (and the shooter must continue to shoot until he either wins or loses), the shooter wins if he rolls his "point" before he rolls a 7. Otherwise he loses.

Example 1. What is the probability that the shooter will win on the first roll? That he will lose on the first roll? That he will neither win nor lose on the first roll?

Solution: The shooter is using two dice, either of which can come up in six different ways. Accordingly, there are $6 \cdot 6 = 36$ different ways these two dice can be thrown. The winning and losing ways to throw the dice on the first roll are shown in Table 9-1. Hence the probability of winning on the first roll is $\frac{8}{36}$ and the probability of losing on the first roll is $\frac{4}{36}$. Thus, on the first roll, the shooter is twice as likely to win as to lose. The probability of neither winning nor losing on the first roll is

$$\frac{36 - (8 + 4)}{36} = \frac{24}{36} = \frac{2}{3}.$$

Next, consider a couple of card games. Of all card games, perhaps the most famous is poker. Draw poker is played by first dealing five cards to each player.

Table 9.1

Ways to Win		Ways to Lose	
Die 1	Die 2	Die 1	Die 2
1	6	1	1
2	5	1	2
3	4	2	1
4	3	6	6
5	2		
6	1		
5	6		
6	5		

After some betting and exchanging of cards and then more betting, the players remaining in the game compare their hands. The possible hands are ranked from highest to lowest. The highest ranking hand wins.

Example 2. A high-ranking poker hand is one that consists of five cards, all of the same suit. Such a hand is called a *flush*. What is the probability of being dealt a flush?

Solution: There are a total of $C(52, 5)$ different five-card poker hands that can be dealt from the standard deck. There are $C(13, 5)$ flushes in any one suit and there are four suits. Hence the probability of being dealt a flush is[3]

$$\frac{4C(13, 5)}{C(52, 5)} = \frac{5148}{2,598,960} \text{ or about } \frac{1}{505}.$$

Example 3. A very fine draw-poker hand is one consisting of three cards all of the same rank and two cards both of another rank. Such hands are called *full houses*. (An example of a full house is the hand 6♣, 6♠, 6◊, 4◊, 4♣.) What is the probability of being dealt a full house?

Solution: Let us first work with a particular full house, the one consisting of three aces and two kings. There are $C(4, 3)$ ways to deal off three aces from the four aces in the deck and there are $C(4, 2)$ ways to deal off two kings from the four kings in the deck. Hence there are $C(4, 3) \cdot C(4, 2) = 24$ different ways to deal off a full house consisting of three aces and two kings. More generally, if R_1 and R_2 are any 2 of the 13 ranks, then there are 24 ways to deal off a full house consisting of three cards of rank R_1 and two cards of rank R_2. But we can

---

[3]This count includes both straight and royal flushes. A straight flush consists of five cards in rank order (such as 5, 6, 7, 8, and 9) all of the same suit. A royal flush consists of 10, J, Q, K, and A all in one suit.

select the rank R_1 in 13 different ways, and after that we can select the rank R_2 in 12 different ways. So there are $24 \cdot 13 \cdot 12 = 3744$ different full houses that can be dealt. Thus the probability of being dealt such a hand is 3744/2,598,960, or about $\frac{1}{694}$.

Because the probability of being dealt a full house is less than the probability of being dealt a flush, full houses rank higher than flushes in draw poker.

Another popular card game is bridge. Thirteen cards are dealt to each of four players and the game begins.

Example 4. What is the probability of being dealt a 13-card hand consisting of all spades?

Solution: The total number of ways to deal 13 cards is $C(52, 13)$. Of all of these hands, only one consists of all spades. Hence the probability of being dealt all spades is

$$\frac{1}{C(52, 13)} = \frac{1}{635,013,559,600}.$$

Example 5. The worst bridge hand that you can be dealt is called a *yarborough*. This is a hand that does not contain any cards higher than a 9. What is the probability of being dealt such a hand?

Solution: There are as many ways to deal a yarborough from a standard deck as there are ways to deal any hand at all from a standard deck from which the 10's, J's, Q's, K's, and A's have been removed. There are 32 cards in such a depleted deck, so there are $C(32, 13)$ different 13-card hands that could be dealt from it. Hence the probability of being dealt a yarborough is

$$\frac{C(32, 13)}{C(52, 13)} = \frac{5394}{9,860,459} \quad \text{or about} \quad \frac{1}{1828}.$$

Keno is a form of lottery and is played in many different ways. One way to play the game is for you to select 8 numbers of your choice from the 80 numbers 1 through 80. After you have made your selection, the house selects 20 numbers at random. You win according to how many of your 8 numbers are included in the 20 numbers chosen by the house. The popularity of this game is a result of the fact that it is possible to win as much as $25,000 on as small a bet as $1.20.

Example 6. You select 8 numbers from the numbers 1 through 80. Then the house selects 20 numbers at random. What is the probability that all 8 of your numbers are chosen by the house?

Solution: There are a total of $C(80, 20)$ ways for the house to choose their 20 numbers. How many of these will include your 8 numbers? Because your 8 numbers are to be included, we need to find the number of ways of choosing 12 more numbers from the remaining 72. This can be done in $C(72, 12)$ ways. Thus the probability of having all 8 of your numbers chosen by the house is

$$\frac{C(72, 12)}{C(80, 20)} = \frac{51}{11,680,845} \quad \text{or about} \quad \frac{1}{229,036}.$$

All of the problems we have seen so far have been solvable using only the definition of probability. In fact, most problems are difficult enough to require either simplified procedures for computing probabilities or advanced techniques that we are not going to discuss. We shall continue the development of probability theory only so far as the introduction of two theorems that can be used to simplify the computations of probabilities in more complicated problems than the ones we have seen so far.

Theorem.
If the probability that an event E will occur as a result of an experiment is p, then the probability that the event will not occur is $1 - p$.

This theorem is not difficult to believe. It says that, if the probability that you will toss a 5 on one throw of a die is $\frac{1}{6}$, then the probability that you will throw some number other than a 5 is $1 - \frac{1}{6} = \frac{5}{6}$. Here is an example of how this theorem can be used.

Example 7.   A coin is tossed a total of three times. What is the probability that *at least* one head will be thrown?

Solution: Using the theorem, we can compute this probability as follows. Since the probability of throwing no heads at all is only $\frac{1}{8}$ (the only favorable outcome is T-T-T), the probability of throwing at least one head is $1 - \frac{1}{8} = \frac{7}{8}$.

As Example 7 hints, problems involving the key words "at least" are often worked most easily by using this theorem.

A second kind of probability problem that one meets fairly often involves two events occurring in succession. For example, the problem

What is the probability of dealing off the two black aces from a shuffled deck in the first two deals?

is one of this type. The *first* event, E_1, is that of dealing off a black ace. The *second* event, E_2, is that of dealing off a black ace. But note that the probability of event E_2's occurring depends upon whether or not event E_1 has occurred. If event E_1

does occur (with probability $\frac{2}{52}$), then the probability of E_2's occurring is only $\frac{1}{51}$ (since there is only one remaining black ace in the depleted deck). But if event E_1 does not occur, then there are still two black aces in the depleted deck so that the probability of event E_2 would be $\frac{2}{51}$. The point is that the probability of E_2's occurring depends upon whether or not the event E_1 did actually occur or not. The theorem that provides the method for solving such problems is

Theorem.
If an event E_1 can occur with probability p_1 and, assuming that E_1 has actually happened, an event E_2 can then occur with probability p_2, the probability that the two events will occur in succession is the product of p_1 and p_2.

Here are two examples of how this theorem can be used.

Example 8. What is the probability of your shuffling a deck of cards and then dealing the two black aces off the top of the deck?

Solution: The probability of dealing a black ace on the first deal is $\frac{2}{52}$. Assuming that this has happened, in the 51 cards remaining there is only one black ace so that the probability of dealing a black ace on the second deal is only $\frac{1}{51}$. Hence the probability of dealing two black aces on the first two deals is $(\frac{2}{52})(\frac{1}{51}) = \frac{1}{1326}$.

Example 9. Figure 9.4 illustrates a delta region in which a river passing point A divides and redivides as it flows into the sea. If you assume that the currents and eddies in this delta have no effect upon the path of a boat floating from point

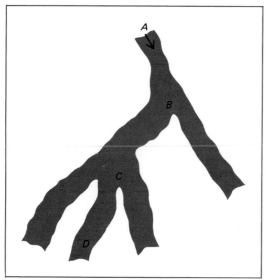

Figure 9.4

A to the sea (an unwarranted assumption that makes this problem much easier to solve), what is the probability that the boat would float from point *A* to point *D*?

Solution: In order for the boat to arrive at point *D*, it would first have to take the left-hand branch at point *B*. (This is the first event—taking the left-hand branch at *B*.) The probability of its doing this is $\frac{1}{2}$. If we assume that this branch has been taken, in order to reach point *D* the boat would next have to take the middle branch at point *C*; the probability that it would do this is $\frac{1}{3}$. Hence the probability that the boat would first take the left-hand branch at point *B* and then the middle branch at point *C* is $(\frac{1}{2})(\frac{1}{3}) = \frac{1}{6}$.

This "boat floating problem" is typical of a large variety of problems called "tree problems." The name was chosen because the diagrams that describe these problems resemble trees if viewed upside down.

Problems of this type can involve more than two events. If such a problem involved three events E_1, E_2, and E_3, then the probability that these three events would occur in succession would be found by computing

1. The probability that event E_1 would occur.
2. The probability that event E_2 would occur, assuming that event E_1 had already occurred.
3. The probability that event E_3 would occur, assuming that event E_1 and event E_2 had occurred.

Then these three probabilities must be multiplied together.

Exercises 9.4 _____

1. What is the probability of dealing five spades from a standard deck? What is the probability of dealing five spades from a deck from which there have already been ten cards dealt, four of which were spades?
2. What is the probability of dealing the cards J◊, Q◊, K◊, and A◊ in that order from a shuffled deck? What is the probability of dealing these four cards in any order at all?
3. How many ways are there to deal two cards from a shuffled deck? What is the probability of dealing off a spade and then a heart from the deck? A spade, then a heart, and then a club? A spade, then a heart, and then another spade?
4. If in playing Keno you mark only 1 number from the 80 available before the house draws their 20 numbers, what is the probability that your number will be drawn? If you mark 2 numbers, what is the probability that exactly 1 of your numbers will be drawn? That at least 1 of your numbers will be drawn?
5. What is the probability of being dealt a five-card poker hand consisting of four cards all of the same rank and a fifth card? Compare this probability

with that of being dealt a full house and decide which of the two hands should be the higher ranking.

6. At the University of Kansas a coed has been dealt a bridge hand consisting of A♠, K♠, Q♠, A♡, K♡, Q♡, A◊, K◊, Q◊, J◊, A♣, K♣, and Q♣. (This is a perfect seven no trump hand.) What is the probability of being dealt 13 cards consisting of 4 aces, 4 kings, 4 queens, and 1 jack?

7. We remarked in an earlier section that red had once come up 32 times in a row at Monte Carlo. What is the probability of this event?

8. A coin is tossed five times.
   (a) What is the probability of getting no tails at all?
   (b) What is the probability of getting at least one tail?
   (c) What is the probability of getting at least one head?
   (d) What is the probability of getting at least one head and one tail?

9. From a standard deck two cards are dealt.
   (a) What is the probability that no hearts will be included in these two cards?
   (b) What is the probability that at least one heart will be included? One spade? One diamond? One club?

10. What is the probability of picking up a shuffled deck of cards and dealing off the four aces in succession?

11. The system of channels in the illustration is held vertically and then a marble is dropped into the opening at the top. If you assume that at each branching the marble is as likely to fall to the left as it is to fall to the right, what is the probability that the marble will eventually fall past each of the lettered positions at the bottom of the system?

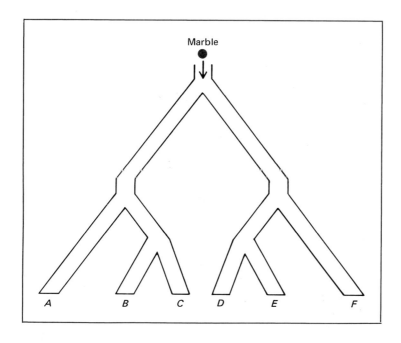

The probabilities that you have obtained should add up to 1 because the possible outcomes of the fall of the marble are exhaustive (the marble must fall past at least one of these points) and mutually exclusive (the marble must fall past at most one of these points).

12. A jug contains five white marbles, six blue marbles, three red marbles and one black marble. Two marbles are to be drawn from this jug.
   (a) If the first marble drawn is blue, what is the probability that the second marble will be blue? Will be red? Will be black?
   (b) If the first marble drawn is white, what is the probability that the second marble will also be white? Will be blue? Will be red?

13. A political hopeful is considering running for mayor of his town. He must win in the primary election and in the general election. He estimates the probability of his winning the primary is about .55 and the probability, assuming he wins in the primary, that he can win the general election is about .6. What is the probability that this candidate can be elected mayor?

14. Russian roulette is played by two players as follows: One bullet is placed into the empty cylinder of a six-gun and the cylinder is spun. The first player holds the gun to his head and pulls the trigger. If he dies, the second player wins automatically. If he draws a blank, the second player takes his turn by spinning the cylinder, putting the gun to his head, and pulling the trigger. If he dies, the first player wins. If he does not die, the game is played again. Would it be better to go first or second in this game? (*Hint:* Compute the probability that the first player will die. Then, assuming that the first player did not die, compute the probability that the second player will die.)

15. Hungarian roulette is played just like Russian roulette except that if the second player has to take a turn, he does not spin the cylinder before he pulls the trigger. Would it be better to go first or second in this game?

16. Bulgarian roulette is played by putting two bullets into the cylinder at random. The first player spins the cylinder, puts the gun to his head, and pulls the trigger. If he dies, the second player wins. If he lives, the second player without spinning the cylinder puts the gun to his head and pulls the trigger. Would it be better to go first or second? Is the situation changed by having the second player spin the cylinder before pulling the trigger?

17. A dart thrower has one chance in three of hitting the bull's eye. What is the probability that he will hit the bull's eye if he throws three darts? (*Hint:* Compute the probability that he will miss on each throw.)

9.5 Mathematical Expectation

One of the important uses of probability is in decision making. In this section we shall introduce the notion of mathematical expectation and point out some of the ways that this idea can be used to aid in making decisions.

Let us marshal the facts that we shall need. Suppose that we are interested in a certain game (such as roulette or craps) which has a finite number of mutually exclusive outcomes represented by $O_1, O_2, \ldots, O_n$. Suppose that the probability of an outcome O_k's occurring is p_k, and that in the event that an outcome O_k does occur you will receive W_k dollars. Then we define the **mathematical expectation** of this game to be the number

$$p_1 W_1 + p_2 W_2 + \cdots + p_n W_n.$$

This number represents the average amount of money you should expect to receive after a sufficiently large number of games have been played. If this number is greater than the amount you must pay in order to play the game, then, in the mathematical sense the game is not a game at all, it is simply a sure-fire way to make money. If the mathematical expectation is less than the amount you must pay in order to play the game, then again, the game really is not a game, it is just a way of throwing money away. The only true games are those for which the mathematical expectation is equal to the cost of playing the game. In such games the amount you win or lose is not predetermined by the a priori probabilities involved, and you will win or lose according to whatever it is that we call "luck." The easiest game to analyze is roulette, so we shall begin with it.

Example 1. If you bet \$1 on a single number and this number is selected, you will win \$35 and your \$1 bet will be returned to you. If the number is not selected, then you lose your \$1 bet. We can express this by saying that if your number is selected you will win \$ $+35$, whereas if your number is not selected you will win \$ -1. What is the mathematical expectation of betting the number 15?

Solution: The probability of the number 15's being chosen is $\frac{1}{38}$, so the probability that it will not be chosen is $\frac{37}{38}$. You win \$ $+35$ in the first event and \$ -1 in the second. Hence the mathematical expectation of this wager is given by the expression

$$(\tfrac{1}{38})(35) + (\tfrac{37}{38})(-1),$$

which is equal to $\frac{-2}{38}$. Thus you should expect to *lose* \$ $\frac{2}{38}$ or about $5\frac{5}{19}$ cents each time you make this wager.

In Example 1 we say that the house has an *edge* of $5\frac{5}{19}$ per cent on this particular bet. It can be shown that, with one exception (see Exercise 1), every bet that you can make on the roulette table (and there are a great many different kinds of bets possible) has the same expectation. Thus, no matter how you place your bets at roulette (with the one exception), you should expect to lose about $5\frac{5}{19}$ cents for each \$1 that you bet. Since there are other games that give the house less of an edge, roulette is one of the worst gambling games.

It should be clearly understood that when we say you will lose $5\frac{5}{19}$ cents for each dollar that you wager, we mean that if you were to play roulette over a very long period of time, then you would tend to lose money at this rate. The law of large numbers guarantees this and so is working against you at roulette. By the same token the law of large numbers is working for the house; thus, although the house may suffer heavy losses over short periods of time, in the long run they will tend to profit by $5\frac{5}{19}$ cents for every dollar wagered on their tables. In any game in which the mathematical expectation is less than the amount of money you must pay in order to play the game, the law of large numbers works against you and in favor of the house.

You have probably heard people talk about "systems" for beating roulette. Any system that requires continued play over a long period of time must buck the law of large numbers and the house's edge. Since the law of large numbers is immutable, no roulette system that requires extended play is possible. The only way to avoid the consequences of the law of large numbers is to play for so short a time that the law of large numbers does not get a chance to begin working against you. One system of this kind is called the "Bold Strategy" or the "Big Bang Strategy." To gamble on roulette using this strategy, simply walk up to the table and bet everything you want to gamble on a single number. Win or lose, you then leave the table and do not continue to play.

Next, let us examine one of the more popular bets on the craps table. This is the *field bet*. The field is the set of numbers 2, 3, 4, 9, 10, 11, and 12. You can make this bet at any time during the play and if one of these seven field numbers comes upon the next throw of the two dice, then you win. Suppose you bet \$1 on the field. If either a 2 or a 12 is thrown, then you win \$+2. If one of the other numbers in the field is thrown, then you win \$+1. If none of the field numbers is thrown, then you "win" \$−1; that is, you lose your bet.

Example 2. What is the mathematical expectation of a field bet of \$1?

Solution: If on one roll of the dice any one of the numbers 2, 3, 4, 9, 10, 11, or 12 results, then you win. If any other number comes up, you lose. The probabilities of these outcomes are given in the Table 9.2. Therefore, the mathematical expectation of the field bet is

$$\tfrac{1}{36}(2) + \tfrac{2}{36}(1) + \tfrac{3}{36}(1) + \tfrac{4}{36}(1) + \tfrac{3}{36}(1) + \tfrac{2}{36}(1) + \tfrac{1}{36}(2) + \tfrac{20}{36}(-1)$$

or $\frac{-2}{36}$. This means that you should expect (on the basis of the a priori probabilities involved) to lose $\$\frac{2}{36}$ or $5\frac{5}{9}$ cents each time you bet \$1. This is an even worse bet than the roulette game.

Next, let's look at the lottery game Keno. This is a real sucker game. One way to play the game (a way that is relatively easy to work with here) is to pick in any way that you desire 5 numbers from the total of 80. (Such a selection is called a "five spot.") Then the house selects 20 numbers at random. Let us suppose that

Table 9.2

Outcome	Ways Can Occur on the Two Dice	Probability of Occurring	Your Win
2	1–1	$\frac{1}{36}$	+2
3	1–2, 2–1	$\frac{2}{36}$	+1
4	2–2, 1–3, 3–1	$\frac{3}{36}$	+1
9	4–5, 5–4, 3–6, 6–3	$\frac{4}{36}$	+1
10	5–5, 4–6, 6–4	$\frac{3}{36}$	+1
11	5–6, 6–5	$\frac{2}{36}$	+1
12	6–6	$\frac{1}{36}$	+2
Nonfield number	—	$\frac{20}{36}$	−1

you have paid \$28.20 for the privilege of selecting your 5 numbers. If all of your 5 numbers are included in the 20 numbers chosen by the house, then you will win \$25,000.[4]

Example 3. How much should you expect to win or lose by betting \$28.20 for a five-spot Keno selection?

Solution: We must compute the probability of having your 5 numbers chosen by the house. There are $C(80, 20)$ ways for the house to choose their 20 numbers. Starting with your 5 numbers, we can fill out a selection of 20 numbers by drawing 15 numbers from the 75 that remain from the original 80 after you have selected your 5. Hence there are $C(75, 15)$ ways to fill out your 5 numbers to a total of 20 numbers. The probability of having all 5 of your numbers chosen by the house, therefore, is, $C(75, 15)/C(80, 20)$ or about $\frac{1}{1503}$. Each time this happens, you receive \$25,000. (Otherwise you receive \$$-28.20$; that is, you lose \$28.20.) Thus the mathematical expectation of this bet is

$$\left(\frac{1}{1503}\right)(25000) + \left(\frac{1502}{1503}\right)(-28.20) = \$-11.54 \qquad \text{approximately.}$$

Therefore, you should expect to lose \$11.54 each time you make this bet. This works out to a house edge of over 40 per cent.[5]

In our discussion of mathematical probability as it applies to games of chance, all the probabilities were definitely established a priori probabilities and all of the payoffs were definite. In other less exotic situations to which mathematical expectation might be applied to help one decide on the best course of action,

---

[4]The bet of \$28.20 is the smallest bet you can make on a five-spot ticket and win the limit, \$25,000.

[5]If only 3 of your numbers are drawn, you will win \$23.50, and if 4 numbers are drawn, you will win \$235. (This works out to a mathematical expectation of \$20.83 per game and means a per game loss of \$7.27 and a house edge of 26 per cent.) We ignored these smaller winnings in order to simplify the computations.

Table 9.3

	Biergarten	Brauhaus
Rain ($\frac{2}{5}$)	$50	$500
Sun ($\frac{3}{5}$)	$450	$100

neither the probabilities nor the payoffs are likely to be so definitely established. Here is an example of one of these more ordinary problems in which mathematical expectation can be used to aid in decision making.

Example 4. A man in Seattle wants to open either a Brauhaus or a Biergarten. He knows that it rains a good deal in Seattle and that on rainy days the Biergarten will not do nearly as good a business as the Brauhaus. But on sunny days the Biergarten will do a fine business. He estimates that the probability of rain on the average day in Seattle is $\frac{2}{5}$. Further, he estimates that he would receive net profits according to Table 9.3.

The dollar amounts in this table represent net profit. For example, the net profit from the Brauhaus on a rainy day would be $500. Should the man open a Brauhaus or a Biergarten?

Solution: Before reading this solution, you might like to come to a decision for yourself based upon the evidence just presented.

Let's apply mathematical expectation and see what we might conclude. The net profits play the role of the payoff rates in our discussions of gambling games. Hence the mathematical expectation of the Biergarten is

$$\tfrac{2}{5}(50) + \tfrac{3}{5}(450) = 20 + 270 = 290,$$

whereas the mathematical expectation of the Brauhaus is

$$\tfrac{2}{5}(500) + \tfrac{3}{5}(100) = 200 + 60 = 260.$$

We conclude, therefore, that we can expect a profit of $290 a day from the Biergarten and a $30 smaller daily profit from the Brauhaus. Hence, according to this analysis, the man should definitely open the Biergarten. If he stays in business a long time, he will tend to make more money with the Biergarten than with the Brauhaus.

Exercises 9.5 ————————————————————

1. We remarked in the text that all bets on roulette except one yield the same house percentage. This exception is the bet (placed on the lower left-hand corner

of the square containing the number 1) on the set of numbers 0, 00, 1, 2, and 3. If you bet $1 on this combination you will win $+6 if one of these numbers is selected and you will "win" $-1 if none of them is selected. Is this a better bet than the others or is it worse?

2. If you place $1 on the square marked "First 12," you are betting that one of the numbers 1 through 12 will come up. If this happens, then you will win $+2. If it doesn't, then you will "win" $-1. Show that the mathematical expectation for this wager is the same as betting on a single number.

3. If you place $1 on the square marked "2 to 1" (next to the square containing the number 34), then you are betting that one of the numbers 1, 4, 7, 10, ..., 31, or 34 will come up. If one of these numbers does come up, then you will win $+2; otherwise you "win" $-1. Show that the mathematical expectation for this bet is the same as that for a single number bet.

4. Suppose that the winner of a lottery will receive a prize worth $500. Each ticket is to cost $1. How many tickets should be sold in order that the lottery be completely fair?

5. You toss a die and are paid an amount in dollars equal to the number you have tossed. What is the minimum amount of money you should expect to have to pay in order to play this game?

6. *Chuck-a-Luck* is a game played with three dice. You play by placing your bet on any one of the numbers 1 through 6. The three dice are thrown and you are paid off as follows:

> If your number comes up on exactly one die: $+1.
> If your number comes up on exactly two dice: $+2.
> If your number comes up on exactly three dice: $+3.
> If your number does not come up on any die: $-1.

How much should you expect to lose every time you bet $1 on the number 1? How does Chuck-a-Luck compare with roulette, Keno (betting 8 numbers), and the field bet at craps?

7. Why can't Chuck-a-Luck be "beaten" by betting $1 on each of the numbers 1 through 6? That is, you will lose money even if you bet $1 on each of the six numbers. Why?

8. A man buys an item costing $6. He tells the salesgirl that he has four $1 bills in his pocket together with two $5 bills and two $10 bills. He offers to reach into his pocket and pull out two bills at random and give them to the girl in return for the item he is buying. Should the salesgirl consider this proposition?

9. The kingdoms of Klutz and Smorg are engaged in an arms race, the cost of which is getting out of hand. The king of Klutz has placed "desirability ratings" upon various contingencies connected with the disarming of the Kingdom of Klutz. The king figures that the desirability factor of disarming is −10 if Smorg invades as a consequence of the disarming and is +10 if Smorg does

not invade. On the other hand, the desirability factor of continuing the arms race is $+5$ if Smorg has plans to invade Klutz and is $+2$ if Smorg does not intend to invade. If the chances of Smorg's invading Klutz are 40 per cent, should Klutz arm or disarm? If this percentage is 30, what should Klutz do?

10

A Glimpse at Statistics

When some people use the word statistics, they are referring to lists of numbers such as the lengths of bridges or the populations of countries. However, there is a much larger meaning for the word, and we shall study some aspects of this larger meaning in this chapter. Statistics has been called "the science and art of making wise decisions in the face of uncertainty." Section 10.6 illustrates one technique by means of which the statistician goes about doing this. The first five sections introduce basic notions.

10.1 Introduction

If we need to give a beginning for the mathematical study of statistics, we should begin in the year 1763 with an English clergyman, Thomas Bayes, and a theorem that today bears his name. We shall not try to state Bayes' Theorem (it is too technical), but we can discuss what this theorem enables us to do. In order to make our discussion concrete, put yourself in the position of a meat processor who would like to learn more about the meat-eating habits of Americans. Because you have considerable experience in the processing and selling of meat products, you have already formed some conclusions about these meat-eating habits; but you would like to confirm and refine your conclusions by actually interviewing people and asking them questions at first hand.

Obviously, although it might be nice to interview each and every American, this is hardly possible. So, instead of trying to interview everybody, you select a small sample of the total population. You intend to interview each member of this sample with the hope that, after you have collected and summarized their responses to your questions, you will be able to use the sample data obtained to modify your original conclusions as to meat-eating habits. This is reasonable and seems to make sense. In fact, Bayes' Theorem provides the mathematical basis for doing precisely this. What Bayes' Theorem says is that there exists a mathematical way to use the sample data to modify your previously drawn conclusions as to the meat-eating habits of the entire population. One way to think of it is that Bayes' Theorem gives mathematical credence to our intuitive belief that we ought to be able to gain information about a total population by drawing off a sample of that population and studying only the members of the sample.

We can state the central problem of statistics as follows: What conclusions can we draw, and how confident can we be in those conclusions, concerning a population by studying samples of that population? This general question can be broken down into three slightly more specific questions that statistics attempts to answer:

1. Given a sample of a population, what conclusions about the entire population can be drawn from that sample?
2. How reliable are these conclusions? How confident can we be in them?
3. How can we most intelligently select samples from the population in order to be able to draw the most reliable conclusions?

The first two of these questions suggest that a statistical answer to a problem concerning the probable characteristics of a population will consist of two parts: (1) some conclusions about those characteristics and (2) an indication of how reliable these conclusions are. Thus the statistician will answer a question something like this, "The best conclusion that I can draw from the sample that you have given me is that . . . and the probability that this conclusion is correct is . . . ."

Note, then, that a statistical conclusion is never (well, hardly ever) correct with probability 1. Frequently, statistical conclusions are given either at the 95 per cent or the 99 per cent **confidence level.** This means that the statistical conclusion has either a probability of $\frac{95}{100}$ or $\frac{99}{100}$ of being correct. In a great many applications the 95 per cent confidence level is entirely satisfactory; but often, if the dependability of the conclusion is especially important, the statistical work is done in such a way that the conclusion obtained will have a probability of $\frac{99}{100}$ of being correct.

Question 3 is vitally important. If we do not draw samples intelligently, we are going to arrive at statistical conclusions that are worthless. If you want to determine how the people of Keokuk feel about movies, then you should not hold all your interviews in the lobby of one of the local movie houses. Neither should you wait until 9 P.M. to conduct your interviews by phone.

The best way to take samples of unknown populations is in as random a way as you can devise. Suppose that you wanted to make a telephone survey of the people in a certain town about opinions on fluoridation. There are 10,000 people in the town, and you want to sample 1000 people. You could make a ten-sided die and then, working from the telephone directory, you could go from name to name, tossing the die each time you came to a new name. Whenever the die came up 10, you would call the person whose name you had reached. If the die came up any other number, you would go on to the next person and toss the die again. The a priori probability is that you would in this way select 1000 people to call and your selection would be based on a method as nearly random as anything you could easily devise.

Exercises 10.1 _____

1. List some reasons why it might be absolutely essential to work with only a small sample of a total population rather than with the entire population. (For example, consider the testing of light bulbs to determine their average life span.)

2. You want to determine the movie-going habits of the people in a certain town. To do this, you plan to take a sample of 100 people living in this town and interview them. The problem arises as to how to identify these 100 people. Discuss the advantages and disadvantages of the following techniques:
   (a) Wait until 9 P.M. and then call 100 people chosen randomly from the phone book.
   (b) Stand on the corner of the block containing a movie house and interview the first 100 people you meet.
   (c) Interview 100 high school students.
   (d) Suppose the population of the city is 10,000. Make a 100-sided die and go down the phone book. Each time you come to a name in the book, throw the die. If it comes up "1," make the call; otherwise go on to the next name.

3. The United States Census is an attempt to interview at least one member of every family unit in the country. Do you think that the census is ever successful in accomplishing this?

4. Have you ever been a member of a population sample? Describe any weaknesses or strengths you discerned in the polling procedures used in your case.

5. Give examples of sampling procedures that are used in the United States (for example, the Harris Poll). Do you know of times when such polls made huge errors?

10.2 Histograms

Because statistical tests involve the consideration of numerical data obtained from samples of a large population, we must begin by examining some of the ways that such data can be organized and presented to the user in a coherent and usable form. Let us consider an illustration. Each year for 7 years all entering freshmen at a certain university have been given a comprehensive algebra test. The possible scores on this test are 0 through 10. The teacher of a certain trigonometry class consisting of 14 students obtained the test results for these students. These scores were: 3, 7, 9, 5, 7, 6, 1, 10, 7, 4, 7, 8, 5, and 7. Presented in this way, it is difficult to make much sense from this data. It might be better to at least present these scores in decreasing or increasing order. Better yet, we might prefer to present the data in the form of a table like that shown in Table 10.1. This kind of table is called a **frequency table** because it not only identifies each of the 14 test scores, but also identifies the frequency with which each of these scores occurs.

Before going further, let us introduce some other terms that will have special meaning for us in this chapter. By a **population** we mean the totality of objects that we wish to describe by means of our statistical results. By a **population sample** or simply **sample** we mean the collection of particular objects of the population that we are going to study with the aim of obtaining information that can be used to reveal information about the population itself. The **sample data** is the numerical information we obtain from the objects of the sample, and a **sample datum** is an individual piece of this numerical information. In our illustration, the population is the totality of all freshmen who have taken this algebra test over the years; the sample is the particular trigonometry class; the sample data is the collection of 14 scores; and a sample datum is any one of these individual scores.

Now back to the illustration. The frequency table in Table 10.1 can be rendered pictorially in a useful way by means of what is called a **histogram.** The histogram for Table 10.1 is shown in Figure 10.1.

Table 10.1

Score	Frequency
10	1
9	1
8	1
7	5
6	1
5	2
4	1
3	1
2	0
1	1
0	0

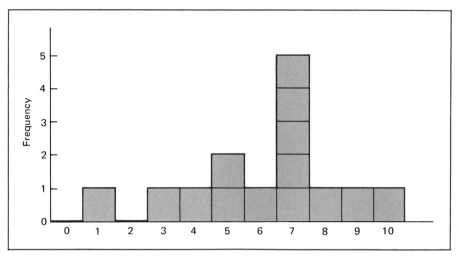

Figure 10.1

The mechanics of constructing this histogram are not complicated. The range of possible test scores is located at equal intervals along the horizontal axis, and the frequencies are located at equal intervals along the vertical axis. A vertical bar over each score is used to indicate the frequency for that score in the sample. These vertical bars are placed abutting each other—there is no space between them. Here are some examples of frequency tables and their histograms.

Example 1. A university selected 100 students with identical programs of study during a certain semester. Each of these students was enrolled in an education course, a mathematics course, a history course, an engineering course, and a sociology course. The letter grades that these 100 students received at the end of the term were collected and summarized by the frequency table shown in Table 10.2. Each column of this frequency table gives rise to a different histogram. The five histograms are shown in Figure 10.2. These histograms provide information only about the particular 100-student sample for which data has been obtained; but, if we assume that the distribution of grades in the sample accurately represents

Table 10.2

Letter Grades	Education	Mathematics	History	Engineering	Sociology
A	58	25	35	28	32
B	28	27	35	35	38
C	9	33	21	30	23
D	3	11	6	6	5
E	2	4	3	1	2

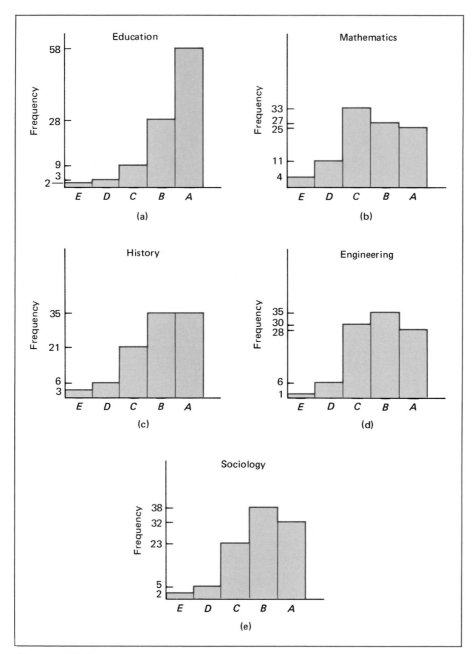

Figure 10.2

the distribution of all student grades, then we may use these histograms to provide information about the total population of all students at this university. For example, we may conclude that the education department grades more easily than do the other departments and that the engineering and mathematics departments have very similar grading standards.

The frequency table and its corresponding histogram contain exactly the same information. We often prefer to present data via the histogram, however, because the geometric presentation of the histogram is often a bit easier to comprehend than the static display of the same data in the form of the frequency table.

Often it suits our purpose to represent a histogram by a smooth curve instead of the broken line curve of a true histogram. In Figure 10.3 the curves in color represent the result of smoothing off each of the histograms in Figure 10.2. The usual procedure used to draw these smoothed-off curves is to connect the midpoints of the horizontal plateaus of the histogram with a smooth curve.

Histograms are nothing more than graphical presentations of frequency tables, and any information obtainable from the table is also obtainable from the histogram, and conversely. Consider this example.

Example 2.   Ten coins were tossed together and the number of heads so obtained was recorded. This experiment was performed 20 times; the results obtained are shown in Table 10.3 and Figure 10.4. Using the frequency table, determine what percentage of tosses resulted in exactly six heads. Determine what percentage of tosses resulted in fewer than six heads and what percentage resulted in more than six heads. Determine the same three percentages using only the histogram.

Solution: Looking at the frequency table, we see that the total number of frequencies is 20 and six was obtained 3 times. Hence the percentage of tosses that resulted in exactly six heads was $\frac{3}{20}$ or 15 per cent. The combined frequencies of tosses resulting in fewer than six heads is $5 + 5 + 2 + 0 + 0 + 0 = 12$ and

Table 10.3

Number of Heads	Frequency
10	0
9	1
8	3
7	1
6	3
5	5
4	5
3	2
2	0
1	0
0	0

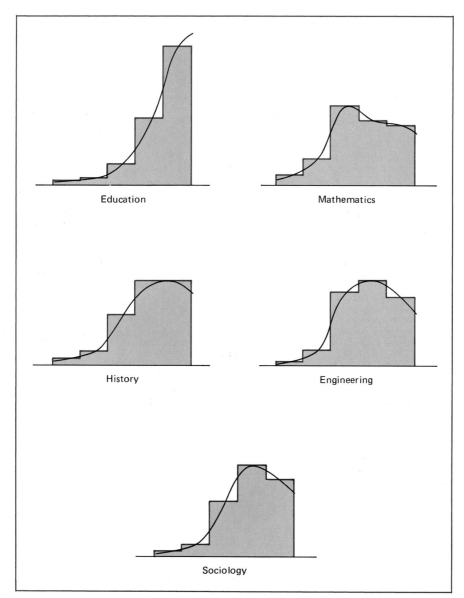

Figure 10.3

so $\frac{12}{20}$ or 60 per cent of the tosses resulted in fewer than six heads. The combined frequencies of the tosses resulting in more than six heads was $1 + 3 + 1 + 0 = 5$ and thus the percentage of tosses resulting in more than six heads was $\frac{5}{20}$ or 25 per cent. (Note that these three percentages sum to 100 per cent as they should—a partial check on our results.)

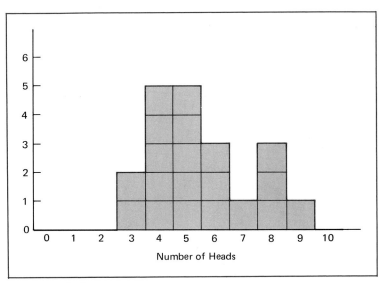

Figure 10.4

Now using the histogram instead of looking at frequencies, we look at the area under the broken line curve. The area of the vertical bar over "6" is three square units and the total area under the curve is 20 square units. Hence $\frac{3}{20}$ or 15 per cent of the total area belongs to the datum "6." This is the geometric way of saying that 15 per cent of the tosses resulted in exactly six heads. The area to the left of "6" is 12 square units, and so $\frac{12}{20}$ or 60 per cent of the area lies to the left of "6"; this is the geometric way of saying that 60 per cent of the tosses resulted in fewer than six heads. Since 25 per cent of the area lies to the right of "6" (five square units of area divided by the total of 20 square units), 25 per cent of the tosses resulted in more than six heads.

Example 2 illustrates that the area under the broken line curve of the histogram can be used in exactly the same way that the frequency can be used in working with the frequency table. The connection between frequency and area that we used in this example is

$$\frac{\text{area of an individual vertical bar}}{\text{total area of the histogram}} = \frac{\text{frequency of the corresponding datum}}{\text{sample size}}.$$

This relationship between the frequency table and the histogram will be put to use in Section 10.5.

So now we have two ways of presenting a collection of sample data—frequency tables and histograms. But we are still faced with the problem of making "sense" from the data. Somehow, the full presentation of the raw data is unsatisfactory—we are not able to cope with it in its raw state. In the next section we shall consider some of the ways that a collection of raw data can be summarized conveniently by a few numerical values.

Exercises 10.2

1. Consider the following frequency tables.

Datum	Frequency
A	0
B	3
C	7
D	4
E	0
F	2
G	3
H	2
I	1

Datum	Frequency
x_1	10
x_2	35
x_3	75
x_4	150
x_5	85
x_6	70

(a) Construct histograms for each of the frequency tables.
(b) Draw smooth curves to approximate these histograms by drawing the curves through the midpoints of the tops of the vertical bars.

2. If the sample size is very large, it may be convenient or even necessary to *group* the data so as to simplify the construction of the histogram. The table at the top of p. 283 lists the number of heads obtained by tossing two pennies 65 times. Construct a histogram by grouping the data in groups of five. That is, consider the first five tosses as comprising the first datum, the second five tosses as comprising the second datum, and so on. The first datum would have a frequency of 2, the second would have a frequency of 4, and so on.

3. Members of an eleventh grade class were weighed to the nearest pound. Their weights were

121	117	137	146
157	95	117	125
137	110	95	117
117	96	119	127
114	113	137	126
104	119	117	147
100	102	126	131
164	164	169	119

(a) Group these weights in 2-pound groups and make a frequency table and a histogram. Draw the smoothed-out curve.

Toss	Number of Heads	Toss	Number of Heads	Toss	Number of Heads
1	0	23	0	45	1
2	0	24	1	46	1
3	0	25	1	47	2
4	0	26	1	48	2
5	2	27	2	49	2
6	0	28	1	50	1
7	2	29	2	51	0
8	0	30	0	52	0
9	0	31	2	53	2
10	2	32	2	54	1
11	2	33	0	55	1
12	0	34	1	56	2
13	0	35	2	57	1
14	1	36	1	58	2
15	1	37	1	59	0
16	0	38	2	60	1
17	2	39	1	61	1
18	1	40	0	62	2
19	2	41	1	63	0
20	1	42	0	64	1
21	0	43	1	65	2
22	1	44	1		

(b) Group these weights in 5-pound groups and make a frequency table and a histogram. Draw the smoothed-out curve.

(c) Group these weights in 10-pound groups and make a frequency table and a histogram. Draw the smoothed-out curve.

4. By choosing appropriate units for the vertical or horizontal scales, we can attempt to mislead people who will be looking at the histogram. Construct a histogram based upon the following data as it might be constructed by the dogfood manufacturer to show dramatic gains in sales. (You would want the columns to rise rapidly rather than slowly.)

Month	Cases of Dog Food Sold	Month	Cases of Dog Food Sold
Jan.	1113	July	1361
Feb.	1235	Aug.	1403
Mar.	1300	Sep.	1417
Apr.	1342	Oct.	1435
May	1345	Nov.	1457
June	1347	Dec.	1479

5. Put yourself in the place of a disgruntled stockholder in the dogfood company of the preceding exercise. Suppose you wanted to downgrade the performance

of the company. How would you make a histogram to reflect the monthly sales if you wanted to deemphasize the steady increase over the year?

6. Get four pennies or other coins and toss the coins together 32 times. Each time you toss the coins, record the number of heads. Make a frequency table and histogram from this experiment. Then, using the same data, make a frequency table and histogram for the number of tails obtained. Compare the histograms. Comment upon this comparison.

7. Construct the frequency table and histogram that you would theoretically obtain by tossing four coins 32 times. That is, you can predict the a priori probability of getting 0 heads, 1 head, 2 heads, and so on. (There are $2 \cdot 2 \cdot 2 \cdot 2 = 16$ ways for the four coins to be tossed. Exactly one of these will result in 0 heads, and exactly four ways will result in 1 head. You carry it from there.) Compare the histogram in Exercise 4 with this one. How well does your experimental histogram "fit" the theoretical one?

10.3 Describing Data Numerically

One way to describe a collection of sample data is to present the complete frequency table or histogram. But this presentation is not concise enough to enable us to make ready sense of the statistical meaning of the data. What is needed is a simpler description of the data. The usual aim is to find two numbers, one of which will identify the "center" of the distribution of the sample data as illustrated by the histogram and the other being used to describe how this distribution of data spreads out and away from this center. This is exactly the same thing as describing Mt. Rainier by specifying its height and the diameter of its base. Much of what is significant about Mt. Rainier is lost by simply presenting these two numbers, but much is retained too. At least these two numbers would enable you to distinguish between Mt. Rainier and Mt. Fuji, or between Mt. Rainier and the compost heap in the author's back yard.

Our first job then is to explain how we will find the "center" of a histogram.

Example 1. Figure 10.5 shows two *smoothed-out* histograms. Histogram (a) has a more or less obvious and natural center. Most of us would agree that M is the center of this population distribution. Histogram (b) presents a problem. Would you want to call M the center of this distribution or would you want to say that this distribution has two centers, A and B? There is no single definitive answer to the question until you know more about the actual problem being studied.

There are a number of ways to identify numerically the center of a sample distribution. By far the most important of these is the **sample mean.** The mean

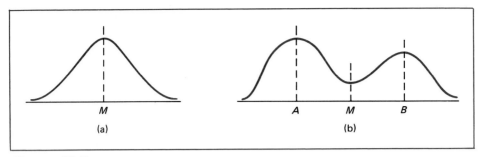

Figure 10.5

of a set of sample data is the average of the sample data; that is, the sum of the sample data divided by the size of the sample. We shall denote the mean of a set of sample data by *m*.

Example 2.　If the ages of a class of students are 18, 18, 20, 19, 17, 17, 19, and 21, then the mean age of this sample of students is

$$m = \frac{18 + 18 + 20 + 19 + 17 + 17 + 19 + 21}{8} = 18.6.$$

The mean will not always be the best measure of the center of the distribution. When the mean is not useful, other *measures of central tendency* may be used. One of these is the **sample mode.** This is nothing more than the sample datum with the greatest frequency. The **sample median** of a set of sample data is that datum with the property that there are as many data above it as there are data below it. If there is no datum with exactly as many data above it as there are data below it, then the midpoint between the two nearest data is taken to be the median. Here is an example illustrating these different measures of central tendency.

Example 3.　For the collection of sample data whose frequency table is Table 10.4, the mode is 12 and the median is 14. Had the datum 11 appeared with a frequency of 2 instead of 3, then the median would have been 14.5 because there would then be 10 data above 14.5 and 10 data below. The mean of this collection of data is about 13.8. These three "centers" are located on the histogram in Figure 10.6.

In the absence of other compelling reasons for using one of the methods of finding the center of a distribution, we use common sense. Consider this example.

Example 4.　Here are final examination scores obtained in an advanced mathematics course: 100, 100, 100, 97, 54, 23, 23, 15, 12, 10, 5, and 2. What kind of a center should we use for this sample distribution?

Table 10.4

Datum	Frequency
11	3
12	6
13	1
14	1
15	5
16	3
17	2

Solution: The mean of these scores is about 45, their mode is 100, and their median is 23. Which of these numbers—100, 45, or 23—should be used to represent the center of this distribution? To use 100 would be to imply that the students got very high grades on this final, whereas actually more than half the students appear to have failed. To use 45 would imply that the class had not done particularly well, even though four of them did superbly. To use 23 as a central score completely misleads. Perhaps none of these central measures is much good here and the only thing to do is to use the entire histogram (or frequency table) to indicate the results.

Once we have some indication of the center of a population distribution, we next must find a way to measure the manner in which the distribution spreads out from that center. There is an elementary way to measure this spread, and it is the same measure of spread as you might use in measuring the "spread" of

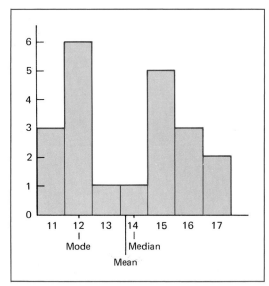

Figure 10.6

Mt. Rainier—you would select two points at the base of the mountain on opposite sides of the center and measure the distance between them. Translated into histograms, this means we subtract the smallest datum from the largest, the difference being the amount of spread. This number is called the **sample range.**

Example 5. In Example 2 the range of ages is the difference between 21 and 17 or 4. In Example 3 the range of the data is from 11 to 17 or 6. In Example 4 the range of test scores is $100 - 2 = 98$.

In some situations the range is adequate, even useful, as a measure of spread. However, in most statistical studies it is not because it is too sensitive to abnormal highs and lows. As a measure of the spread of a distribution about its center, the range can be completely misleading if there is an abnormally small datum or an abnormally large one.

Just as we would have to exercise some judgment as to where the base of Mt. Rainier started, we need a measure of the spread of population distributions that will be able to discern where the significant part of the population begins and ends. The range has no such discernment factor built into it, and so is unusable in most situations. In the next section we shall study another measure of spread that can discern reasonable extremes for the distribution of a population about its center.

Exercises 10.3 ———————————————————————————

1. Find the mean, median, and range of each of the following sets of sample data.
   (a) 60, 73, 56, 74.
   (b) 30, 18, 12, 45.
   (c) 5, −3, 2, −1, 0, −1, 4, 3, −2, −5, 2, 1, 0, 3.
   (d) 1, 2, 3, 4, 5, 6, 7, 8, 9.
   (e) 17, 18, 19.
   (f) 16, 17, 18, 19, 20.
   (g) $x - 2$, $x - 1$, x, $x + 1$, $x + 2$.
2. Construct a histogram for each of the sets of sample data given in Exercise 1 and locate the three "centers" of each distribution.
3. Locate the mode and median of each of these sample data distributions:

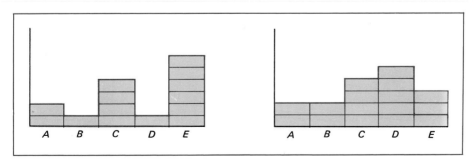

4. The word "average" is often used in very different ways. What would be the effect of using each of the mean, median, and mode to represent the meaning of "average" in each of these statements? Compare the different meanings for appropriateness.
   (a) The average salary of an insurance agent, his secretary, his part-time bookkeeper, and the woman who cleans the office.
   (b) The average number of paper napkins used by a patron of a pizza parlor.
   (c) The average number of miles driven by a vacationer from the state of New York during the month of August in 1971.
   (d) The average size of shoes sold during a given month at a particular store.
5. Ten quizzes are given in a course during the term. After the first five quizzes the mean is computed for purposes of giving midterm grades. After the remaining five quizzes have been given, the mean of the last five quizzes is computed. Then the mean of the first five quizzes and the mean of the last five are added together and divided by 2. Is the number so obtained the same as the mean of all ten scores? Explain as carefully as you can. If your answer is no, then construct an example that will justify your answer.
6. An airport served a mean of 2.3 cups of coffee to each of 1200 travelers caught in the airport during a shutdown of the airport during the winter of 1972. How many cups of coffee were served? Would the management of the airport have any use for the mode of this distribution? That is, would they want to know the largest number of cups of coffee served to any one customer? Would they be interested in the median?
*7. Find the mean, m, of this data:

   15, 18, 21, 14, 12, 24, 18, 16, 17, 14, 13, 15, 24.

   For a given datum, x, its **deviation from the mean** is the difference $x - m$. Find the deviation of each datum from the mean. What is the sum of all 13 deviations? Is the sum you have obtained an isolated oddity or would you always obtain the same sum no matter what data you had?
8. A student received 71, 77, and 81 on four hour-long tests. He must average 80 on five tests to get a B. What score must he shoot for on the last test? (By "average" here we mean the mean. This is the usual meaning of "average.")
9. The class average (that is, the class mean) on a test was 68. After returning the test, the instructor discovered that he had misgraded one of the problems. He decided to give everybody eight more points. What was the corrected mean of this test?
*10. **Coding** is a technique for simplifying the computation of means that is useful in some cases. As an illustration of this technique, compute the mean of the numbers 100, 101, 104, 104, 105, 106, and 108. Then subtract 100 from each number, compute the mean of the resulting differences, and add 100 to this mean. Do you get the mean of the original data? State the theorem involved here. Can you prove this theorem?
*11. Compute the means of the following data populations by the coding technique discussed in Exercise 10.

(a) 2134, 2157, 2131, 2142.

(b) $x + 2, x + 3, x + 4, x + 5, x + 6$.

12. Find four different data collections consisting of five values such that each data collection has a mean of 5.

13. Make a flow chart for finding the median. Insert a decision step involving the question "Is the size of the data collection even?"

10.4 Standard Deviation

In this section we shall discuss the standard deviation of a sample distribution as a means for determining the spread of that distribution about its mean. Let us introduce this concept by means of an example. The population we are going to study is the population of heights of ten-year-old American boys. We have selected a sample of 26 boys and have recorded their heights. This information is contained in the histogram shown in Figure 10.7. Our purpose is to determine a measure for the spread of this sample distribution away from its mean. Hence we ought first to compute the mean. Reading from the histogram,

$$m = \frac{1(57) + 2(56) + 3(55) + 4(54) + 6(53) + 5(52) + 3(51) + 2(50)}{26}$$

$$= \frac{1381}{26}$$

$$= 53 \qquad \text{approximately.}$$

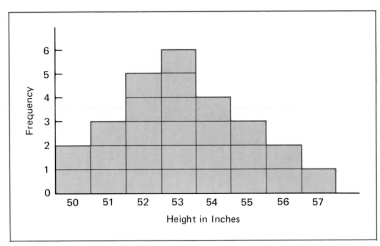

Figure 10.7

Since the spread of the sample data should somehow be related to the way the individual sample datum spreads out from the mean, we next compute what are called the **deviations** of the data from the mean. The deviation of a datum x is the number $x - m$. These deviations appear in the third column of Table 10.5. They tell us how each datum is spread out from the mean either on the left (if the deviation is negative) or on the right (if the deviation is positive).

The next step in computing the standard deviation is a technical one whose justification need not concern us here. We are going to square each of the deviations. These squares are given in the fourth column of Table 10.5.

The next to the last step in computing the standard deviation is to sum the squares of the deviations and divide their sum by *one less than the sample size:*

$$\frac{1(16) + 2(9) + 4(1) + 6(0) + 5(1) + 3(4) + 2(9)}{26 - 1} = \frac{85}{25} = 3.4.$$

The final step is designed to undo the effect of having squared the deviations before we added them—we now take the square root of the number 3.4.[1] The square root of 3.4 is about 1.85. This number is the **sample standard deviation.** We shall denote the sample standard deviation by s.

We can summarize these steps for computing s as follows:

1. Compute m.
2. Compute the deviations.
3. Square the deviations.
4. Add the squares.
5. Divide this sum by one less than the sample size.
6. Take the square root.

Table 10.5

Frequency	Datum	Deviation	Square of Deviation
1	57	4	16
2	56	3	9
3	55	2	4
4	54	1	1
6	53	0	0
5	52	−1	1
3	51	−2	4
2	50	−3	9

[1] You may know some methods for computing square roots, but, if you don't, the simple method of guessing and testing will suffice for our work. First, guess at the square root and then square your guess. If the square is less than the number given, guess again at a larger number; if the square is greater than the number given, guess again at a smaller number. Usually, you can come up with a reasonably precise estimate of the square root in a couple of guesses.

Example 1. Find the standard deviation of the sample data collection consisting of the values 102, 99, 87, 86, 83, 81, 80, 78, 74, 72, 72, 59, and 41.

Solution: We begin by computing the mean: $m = \frac{1014}{13} = 78$. Next we need to find the deviation of each datum from this mean, square these deviations, and sum them. Table 10.6 contains these values. Then we divide the sum of the squares of the deviations by one less than the sample size, 12, and obtain 252. Last, $s = \sqrt{252} = 16$, approximately.

There is a technique that can be used to simplify the computation of the standard deviation in some instances. This device is called **coding.** The basic idea is fairly simple. The smaller the numbers in the sample data collection, the easier it is to compute the standard deviation. Coding is a technique whereby we subtract from each datum a convenient constant. This makes each datum smaller and easier to work with. Here is an example.

Example 2. Use coding to find the standard deviation of 101, 102, 104, 104, and 106.

Solution: We begin by subtracting a convenient constant from each datum in order to make all data as small as possible. In this example we should subtract 100 from each datum. We are then left with the new data collection consisting of 1, 2, 4, 4, and 6. We compute the standard deviation of these data in the usual way. See Table 10.7. The mean of the new data is 3.4. The sum of the squares of the deviations, 15.20, divided by 4, is 3.80; and the square root of 3.80 is about 1.95. The standard deviation of the new data obtained by coding is the same as the standard deviation of the original data. Hence the standard deviation we are seeking is 1.95.

Table 10.6

Datum	Deviations	Squares
102	$102 - 78 = 24$	576
99	$99 - 78 = 21$	441
87	$87 - 78 = 9$	81
86	$86 - 78 = 8$	64
83	$83 - 78 = 5$	25
81	$81 - 78 = 3$	9
80	$80 - 78 = 2$	4
78	$78 - 78 = 0$	0
74	$74 - 78 = -4$	16
72	$72 - 78 = -6$	36
72	$72 - 78 = -6$	36
59	$59 - 78 = -19$	361
41	$41 - 78 = -37$	1369
		3018

Table 10.7

Datum	Deviation	Square
1	2.4	5.76
2	1.4	1.96
4	.6	.36
4	.6	.36
6	2.6	6.76
		15.20

The mean of the new data is equal to the mean of the old data reduced by the code. In Example 2 the mean of the original data is 100 greater than the mean of the coded data: $100 + 3.4 = 103.4$. Refer to Exercise 10 of Section 10.3.

We shall apply the notion of standard deviation in the next two sections.

Exercises 10.4

1. Compute the standard deviation and the mean of each of the following sample data populations:
   (a) 5, 5, 5, 5, 6, 6, 6, 10.
   (b) 1, 2, 3, 4, 5, 6, 7.
   (c) 4, 6, 8, 10, 10, 10, 12, 14, 16.
   (d) 2, 2, 2, 4, 4, 4, 6, 6, 6.
   (e) 6, 6, 7, 7.
   (f) 7, 7, 8, 8.
   (g) -2, 1, 3, -2, -4, 3, 0, 1, 2, -2.
   (h) 4.3, 4.4, 4.2, 4.0, 3.9, 4.1, 3.1.

2. Construct a flow chart showing how to compute the standard deviation of a sample.

3. Compute the mean and standard deviation of the sample data in this frequency table.

Score	Frequency	Score	Frequency
12.0	3	8.0	4
11.5	2	7.9	2
11.0	1	7.8	3
10.5	1	7.5	1
10.0	3	7.0	2
9.5	1	6.5	1
9.0	1	6.0	3
8.6	1	5.5	1
8.4	1	5.0	2
8.3	1	4.5	1
8.2	2	4.0	3
8.1	1		

4. Find an example of a population of five numbers for which the standard deviation is 0. There is a theorem here—can you state it?

5. Two bathroom scales, when tested repeatedly with a weight of 150 pounds, give the following readings:

   Scale A: 150.2, 150.3, 149.5, 150.0, 150.4, 149.6.
   Scale B: 150.1, 149.5, 150.2, 150.1, 149.7, 150.4.

   Which scale is best? (*Hint:* The scale with the smaller standard deviation is best. Coding will make your calculations easier.)

6. Instructors A and B are each scheduled to teach sections of a course you want to take next term. Both instructors have received the same ranking on the faculty teacher ratings that were done by the students last year. However, Instructor A's rankings have a standard deviation of 6, whereas those of Instructor B have a standard deviation of 12. Which instructor would you prefer on the basis of these rankings? Why?

7. Two clocks are tested against an electronic clock for periods of 1 hour. This testing is repeated many times. The mean recorded time for each clock is in fact 1 hour during this 1-hour period. Thus the mean value will not help us to determine whether or not one of these clocks is better than the other. How would you use standard deviations to help you determine the more reliable clock?

8. The formula for the standard deviation requires using the squares of the deviations of the individual data. Explain why it is not possible to use simply the deviations themselves. As an example, compute the standard deviation of 3, 5, 4, 2, 7, and 3, omitting steps 3 and 6 in the computation.

9. Suppose a data collection has a mean of m and a standard deviation of s. Now construct a new data collection by adding 10 to every datum greater than the mean and subtracting 10 from every datum less than the mean. How will the standard deviation of the new data differ from the standard deviation of the original data?

10. Prove that the standard deviation of the numbers x, y, and z is equal to the standard deviation of the numbers $x - 100$, $y - 100$, and $z - 100$.

10.5 Approximately Normal Populations

In the preceding sections we have studied population samples and their numerical characteristics, including what is most important, the sample mean and the sample standard deviation. These two concepts have meaning for entire populations as well. The **population mean,** denoted by μ, is defined to be the sum of the population data divided by the population size. The **population standard deviation,** denoted by σ, is defined exactly as is the sample standard deviation *except,* instead of dividing the sum of the squared deviations by one less than the sample size, we divide by the population size.

For all of the reasons that make sampling procedures necessary, we cannot expect to be able to compute either population means or standard deviations directly according to their definitions. In almost every instance we have to determine these numerical characteristics of entire populations by indirect means that enable us to avoid the enormous or impossible computations that their definitions call for.

The population mean and standard deviation are used just as the sample mean and standard deviation are used. The population mean gives us some idea as to the center of the entire population and the population standard deviation provides information as to the way that the population is distributed around this center. For many populations we would need additional information in order to draw significant statistical conclusions about the nature of the populations, but for some special populations it turns out that the only information we need in order to draw the conclusions we want are the mean and standard deviation of the population. We are going to study such special populations in this section.

Let us begin with an exercise in imagination. Imagine that we have measured the height of each and every American male to the nearest tenth of an inch. The population is the collection of all American males, and the population data consists of their heights. Just as we have done with population samples, we could construct a population histogram that would display this data in a useful form. We could then smooth off this histogram, obtaining a smooth curve that would approximate the histogram. It would turn out that the curve we would obtain would closely resemble the curve shown in Figure 10.8. This particular curve is called a **normal** curve; and, because the smoothed-out population histogram gives rise to such a curve, the population itself is called **approximately normal.**

All normal curves share the general characteristics of the curve in Figure 10.8. Such curves are symmetric with respect to their vertical axis and tail off in both directions, dropping toward the horizontal axis. The left- and right-hand tails never

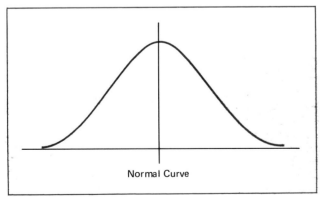

Normal Curve

Figure 10.8

actually reach the horizontal axis, but they continue to get closer and closer to this axis as they move out to the left and to the right. It is not easy to identify a normal curve simply by looking at its graph (for there are other curves that are very similar), so we shall always identify normal curves for you.

Approximately normal populations arise naturally in our study of the real world. Examples of such populations are

1. The weights of Coca Cola bottles measured to the nearest tenth of an ounce.
2. The amounts of garbage collected from American homes in 1974 measured to the nearest pound.
3. The breaking strengths of aircraft cables produced at a certain factory measured to the nearest pound.
4. The resistances of transistors whose rated resistance is 12 ohms measured to the nearest .01 ohm.
5. The weights of newborn mice used in a certain laboratory for psychological experiments measured to the nearest ounce.

A common trait of approximately normal populations is that they are almost always very large. But not all large populations are approximately normal. The population of grades earned by university students enrolled in honors programs in mathematics would probably not be approximately normal. In fact, the smoothed-out histogram for this population would probably look something like the curve of Figure 10.9.

We have said that approximately normal populations are easier to deal with than many others because all we need to know about such a population is its mean

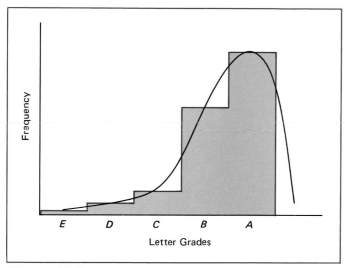

Figure 10.9

and standard deviation. Let us see why this is so. Certainly, any statistical conclusion that we might expect to draw about a population will be deduced in some way through an examination of data. Since the population histogram reveals the population data, any such conclusion could be drawn from an examination of this histogram. But the histogram for an approximately normal population is closely approximated by a normal curve. It follows from this that, with only a small error resulting from the approximation, we can draw the conclusion from an examination of the normal curve associated with the population. All that we need now is the following.

Theorem.
The normal curve associated with an approximately normal population is determined by the mean and standard deviation of that population.

The theorem then tells us that any conclusion we might expect to draw from the raw population data can be drawn from an examination of merely two numbers—the population mean and standard deviation. In a manner of speaking, for this particular class of populations, we are able to distill all of the raw population data into the population mean and standard deviation.

So our job now is twofold. First, we must study normal curves to learn the significance of the population mean and the standard deviation. Second, we must determine how to apply what we have learned about normal curves so that we can draw statistical conclusions. Before we do this, however, we should consider briefly the danger in replacing the true histogram by a normal curve. We can legitimately do this only if the histogram does closely resemble a normal curve—we ought not to force such a replacement simply because of the relative simplicity in dealing with approximately normal populations. However, in order to simplify the statistical work, those who use statistical methods may make unwarranted assumptions to the effect that the population under study is approximately normal. The motivation for making such an assumption is strong; but if the assumption is made on weak grounds, then the resulting conclusions will of course be open to question.

It is not difficult to make a rough sketch of the normal curve associated with an approximately normal population with mean μ and standard deviation σ. Since a normal curve is symmetric about its center, the mean μ must be located there. (Refer to Figure 10.10.) We may select any point P on the vertical axis as the "top" of the normal curve, since we are free to choose the unit distance on the vertical axis in such a way that P would correspond to the frequency with which the mean appears in the population data.

Now the way that the standard deviation enters into the construction of this normal curve is not so obvious. It can be shown that the interval extending three standard deviations on either side of the mean contains over 99 per cent of the total area under the normal curve. So, if we measure off a distance on either side of μ equal to 3σ, we can draw the normal curve as almost touching the horizontal axis at the points $\mu - 3\sigma$ and $\mu + 3\sigma$. (We shall ignore that portion of the curve

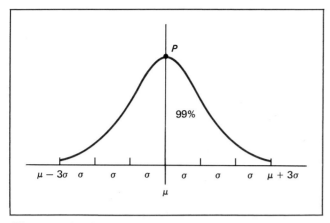

Figure 10.10

that extends further out from the mean than three standard deviations.) Now draw a smooth curve extending downward from P to the points $\mu - 3\sigma$ and $\mu + 3\sigma$ on the horizontal axis. Anything you sketch that looks like the curve in Figure 10.10 will be good enough. The exact shape of this curve is known, but it is difficult to describe accurately without some understanding of calculus.

Example 1. An approximately normal population has a mean of 15 and a standard deviation of 1. Sketch the associated normal curve. What would the associated normal curve look like if the population standard deviation were 2 instead of 1? What if $\sigma = \frac{1}{2}$?

Solution: See Figure 10.11. These curves show that the location of the normal curve on the horizontal axis is determined by μ and that the quickness with which

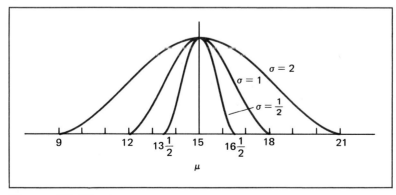

Figure 10.11

the normal curve drops to the horizontal axis is determined by the standard deviation. The smaller the standard deviation, the more quickly the normal curve drops.

We have said that the normal curve associated with an approximately normal population is determined by the mean and standard deviation of that population. In particular, questions relating to the area under the normal curve can be answered purely in terms of μ and σ. In Figure 10.12 we show a normal curve and a point, x, lying to the left of the mean, μ. (If x were to lie to the right of the mean, we could make the same arguments that we are going to make now, but for the time being we shall consider only points to the left of μ.) If we can determine what percentage of the area under the normal curve is in the shaded region to the left of the point x, then we can answer many problems concerning approximately normal populations. Here is an example of such a problem.

Example 2.   A bottle manufacturer produces a bottle which, by means of routine sampling procedures at the end of the production line, is known to have a mean bursting strength of 150.06 pounds per square inch (psi). We assume the population consisting of the bursting strengths of the bottles produced is approximately normal. The manufacturer has determined that the standard deviation of this population is 11.73. What is the probability that a bottle selected at random from the production line will have a bursting strength below 125 psi?

Solution: Consider Figure 10.13. We illustrate here the normally distributed bursting strength of bottles, the mean bursting strength of 150.06 psi, and the bursting strength of 125 psi. Observe that if we knew the area of the shaded portion of the region under the curve, then we would know what percentage of the bottles produced had bursting strengths of less than 125 psi. For example, if this area amounted to 15 per cent of the total area under the normal curve, then we could say that, on the average, 15 out of 100 bottles will burst at pressures under 125 psi. This might be an important piece of information if the bottles were being considered for use in containing a carbonated liquid.

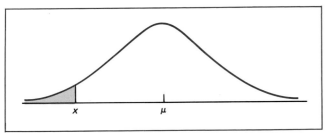

Figure 10.12

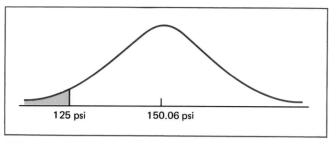

Figure 10.13

In a moment we shall consider additional examples of how knowing the area under a normal curve can provide useful information. But let us first see how to determine the areas of regions under the normal curve. Remember that the area to the left of the point x in Figure 10.12 must be derivable from the mean and the standard deviation of the normal curve. We begin by expressing this point x in terms of these two quantities. First, the distance between x and μ is $\mu - x$. (This is because x lies to the left of μ. If x were to the right, this distance would be $x - \mu$.) Next, by solving the equation

$$\mu - x = k\sigma,$$

where σ is the standard deviation of this curve, we can determine of how many standard deviations this distance is composed. The multiplier k is all important. Using k and Table 10.8, we can determine what percentage of the total area under the curve lies to the left of the point x. Table 10.8 is not difficult to use. For example, if k were 1.3, run down the multiplier column until you come to 1.3. Go across to the corresponding entry in the percentage area column. This percentage is 9.68. We then know that 9.68 per cent of the total area under the curve lies to the left of the point x. Here is a complete example of this simple computation.

Example 3. Suppose that we have an approximately normal population with $\mu = 16.5$ and $\sigma = 1.4$. (See Figure 10.14.) We want to know what percentage of

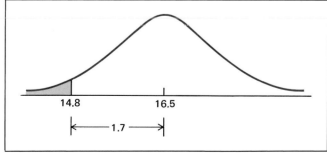

Figure 10.14

Table 10.8 Table of Multipliers and Percentages of Area

k	Per Cent	k	Per Cent	k	Per Cent	k	Per Cent	k	Per Cent	k	Per Cent
4	.003	2.64	.41	2.09	1.83	1.54	6.18	.99	16.11	.44	33.00
3.9	.005	2.63	.43	2.08	1.83	1.53	6.30	.98	16.35	.43	33.36
3.8	.01	2.62	.44	2.07	1.92	1.52	6.43	.97	16.60	.42	33.72
3.7	.01	2.61	.45	2.06	1.97	1.51	6.55	.96	16.85	.41	34.09
3.6	.02	2.60	.47	2.05	2.02	1.50	6.68	.95	17.11	.40	34.46
3.5	.02	2.59	.48	2.04	2.07	1.49	6.81	.94	17.38	.39	34.83
3.4	.03	2.58	.49	2.03	2.12	1.48	6.94	.93	17.62	.38	35.20
3.3	.05	2.57	.51	2.02	2.17	1.47	7.08	.92	17.88	.37	35.57
3.2	.07	2.56	.52	2.01	2.22	1.46	7.22	.91	18.14	.36	35.94
3.1	.10	2.55	.54	2.00	2.28	1.45	7.35	.90	18.41	.35	36.32
3.09	.10	2.54	.55	1.99	2.33	1.44	7.49	.89	18.67	.34	36.69
3.08	.10	2.53	.57	1.98	2.39	1.43	7.64	.88	18.94	.33	37.07
3.07	.11	2.52	.59	1.97	2.44	1.42	7.78	.87	19.22	.32	37.45
3.06	.11	2.51	.60	1.96	2.50	1.41	7.93	.86	19.49	.31	37.83
3.05	.11	2.50	.62	1.95	2.56	1.40	8.08	.85	19.77	.30	38.20
3.04	.12	2.49	.64	1.94	2.62	1.39	8.23	.84	20.05	.29	38.59
3.03	.12	2.48	.68	1.93	2.68	1.38	8.38	.83	20.33	.28	38.97
3.02	.13	2.47	.68	1.92	2.74	1.37	8.53	.82	20.81	.27	39.36
3.01	.13	2.46	.69	1.91	2.81	1.36	8.69	.81	20.90	.26	39.74
3.00	.13	2.45	.71	1.90	2.87	1.35	8.85	.80	21.19	.25	40.13
2.99	.14	2.44	.73	1.89	2.94	1.34	9.01	.79	21.48	.24	40.52
2.98	.14	2.43	.75	1.88	3.01	1.33	9.18	.78	21.77	.23	40.90
2.97	.15	2.42	.78	1.87	3.07	1.32	9.34	.77	22.06	.22	41.29
2.96	.15	2.41	.80	1.86	3.14	1.31	9.51	.76	22.36	.21	41.68
2.95	.16	2.40	.82	1.85	3.22	1.30	9.68	.75	22.66	.20	42.07
2.94	.16	2.39	.84	1.84	3.29	1.29	9.85	.74	22.96	.19	42.47
2.93	.17	2.38	.87	1.83	3.36	1.28	10.03	.73	23.27	.18	42.86
2.92	.17	2.37	.89	1.82	3.44	1.27	10.20	.72	23.58	.17	43.25
2.91	.18	2.36	.91	1.81	3.52	1.26	10.38	.71	23.89	.16	43.64
2.90	.19	2.35	.94	1.80	3.59	1.25	10.56	.70	24.20	.15	44.04

2.89	.19	2.34	.96	1.79	3.67	1.24	10.75	.69	24.51	.14	44.43
2.88	.20	2.33	.99	1.78	3.75	1.23	10.93	.68	24.83	.13	44.83
2.87	.21	2.32	1.02	1.77	3.84	1.22	11.12	.67	25.14	.12	45.22
2.86	.21	2.31	1.04	1.76	3.92	1.21	11.31	.66	25.46	.11	45.62
2.85	.22	2.30	1.07	1.75	4.01	1.20	11.51	.65	25.78	.10	46.02
2.84	.23	2.29	1.10	1.74	4.09	1.19	11.70	.64	26.11	.09	46.41
2.83	.23	2.28	1.13	1.73	4.18	1.18	11.90	.63	26.43	.08	46.81
2.82	.24	2.27	1.16	1.72	4.27	1.17	12.10	.62	26.76	.07	47.21
2.81	.25	2.26	1.19	1.71	4.36	1.16	12.30	.61	27.09	.08	47.61
2.80	.26	2.25	1.22	1.70	4.46	1.15	12.51	.60	27.43	.05	48.01
2.79	.26	2.24	1.25	1.69	4.55	1.14	12.71	.59	27.76	.04	48.40
2.76	.27	2.23	1.29	1.68	4.65	1.13	12.92	.58	28.10	.03	48.80
2.77	.28	2.22	1.32	1.67	4.75	1.12	13.14	.57	28.43	.02	49.20
2.76	.29	2.21	1.36	1.66	4.85	1.11	13.35	.56	28.77	.01	49.60
2.75	.30	2.20	1.39	1.65	4.95	1.10	13.57	.55	29.12	.00	50.00
2.74	.31	2.19	1.43	1.64	5.05	1.09	13.79	.54	29.46		
2.73	.32	2.18	1.46	1.63	5.16	1.08	14.01	.53	29.81		
2.72	.33	2.17	1.50	1.62	5.26	1.07	14.23	.52	30.15		
2.71	.34	2.16	1.54	1.61	5.37	1.06	14.46	.51	30.50		
2.70	.35	2.15	1.58	1.60	5.48	1.05	14.69	.50	30.85		
2.69	.36	2.14	1.62	1.59	5.59	1.04	14.92	.49	31.21		
2.68	.37	2.13	1.66	1.58	5.71	1.03	15.15	.48	31.56		
2.67	.38	2.12	1.70	1.57	5.82	1.02	15.39	.47	31.92		
2.66	.39	2.11	1.74	1.56	5.94	1.01	15.62	.46	32.28		
2.65	.40	2.10	1.79	1.55	6.06	1.00	15.87	.45	32.64		

the total area lies to the left of the point 14.8. We begin by determining the distance between the mean and the point we are concerned with: $\mu - x = 16.5 - 14.8 = 1.7$. Now we determine the number of standard deviations in 1.7. That is, we solve the equation

$$1.7 = k(1.4)$$

in order to determine the multiplier k. The solution is $k = 1.2$, approximately. Now, referring to Table 10.8, we see that corresponding to the multiplier of 1.2 we cut off an area amounting to 11.51 percent of the total area.

So the procedure for determining the percentage of area cut off by a point lying to the left of the mean is

1. Determine the difference between the value of the point and the mean.
2. Divide this difference by the standard deviation to get the multiplier k.
3. Use Table 10.8 to find the percentage area corresponding to the multiplier k.

Now suppose we want to find the area to the *right* of the point x. Then we would first find the percentage of area to the left of x and subtract this from 100 per cent. But we still have the problem of what to do if the point lies to the *right* of the mean. In this event we use the symmetry of the normal curve about its mean to reflect the whole problem into the one we already know how to solve. For example, suppose we want to find the percentage of area to the right of the point x in Figure 10.15. Since the normal curve is symmetric about its mean, this is exactly the same problem as finding the percentage of area to the left of the point x in Figure 10.16 and we already know how to find this area. Thus, given any point under the normal curve, we can find the percentage of the total area that lies either to the left or to the right of that point.

We can also determine percentage of areas of regions that have both ends cut off. The percentage of the area for the region in Figure 10.17, for example, can

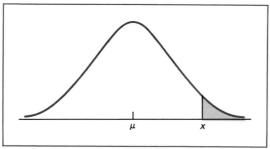

Figure 10.15

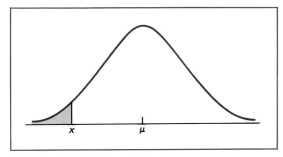

Figure 10.16

be determined by first finding the percentage of the area of the left-hand piece that is cut off by x, then finding the percentage of area of the right-hand piece that is cut off by y, and finally subtracting the sum of these two percentages from 100 per cent.

Example 4. What are the endpoints of the interval, centered on the mean, that cuts off 95 per cent of the total area under a normal curve?

Solution: We want to find x and y in Figure 10.17 so that each of the shaded areas contains 2.5 per cent of the total area under the curve. According to Table 10.8, 1.96 standard deviations to the left of the mean will cut off a region containing 2.5 per cent of the area. So, if we let σ denote the standard deviation, the interval from $x = \mu - (1.96)\sigma$ to $y = \mu + (1.96)\sigma$ determines 95 per cent of the total area under the normal curve.

Example 5. Returning to the bottle manufacturer of Example 2, if 110 psi were regarded as the minimum safety standard for bursting strength, what percentage of the bottles produced would be unsafe?

Solution: All we have to do is to repeat Example 2 using 110 in place of 125. The mean bursting strength is $\mu = 150.06$, and the standard deviation of the

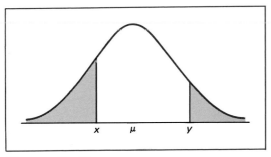

Figure 10.17

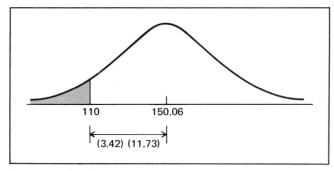

Figure 10.18

normal distribution of bottles produced is $\sigma = 11.73$. (Refer to Figure 10.18.) The point 110 psi is located to the left of the mean a distance equal to $k\sigma$, where

$$k = \frac{150.05 - 110}{11.73} = 3.42 \qquad \text{approximately.}$$

According to Table 10.8, this number of standard deviations to the left of the mean cuts off .03 per cent of the total area. Hence only 3 in 10,000 bottles are likely to burst at an internal pressure of 110 psi. Because it is possible to lose an eye from an exploding soft drink bottle, this would be unacceptable from a consumer point of view.

Example 6. What should be the mean bursting strength of a run of bottles if 125 psi is to be the minimum safe bursting strength, and the percentage of unsafe bottles is expected to be less than .1 per cent (1 in 1000)? We shall assume that the bursting strengths of bottles in this run are normally distributed with a standard deviation of 11.73.

Solution: We want to cut off .1 per cent of the area on the left end under the normal curve with a value of 125 psi. According to Table 10.8 we can do this if 125 psi is located 3.08 standard deviations to the left of the mean. Since 3.08 standard deviations are $(3.08)(11.73) = 36.13$, approximately, 125 should be 36.13 units from the mean. Thus the mean is $125 + 36.13 = 161.13$. So, if a run of bottles is produced whose bursting strengths are normally distributed with a mean bursting strength of 161.13 psi, then fewer than 1 in 1000 bottles will burst at 125 psi.

These examples demonstrate that the connection between standard deviation and percentage area under the normal curve has important applications. The following exercises continue to develop these applications.

Exercises 10.5

1. In an approximately normal population with mean μ and standard deviation σ,
   (a) What percentage of the population is greater than μ? What is the probability that a datum is greater than μ?
   (b) What percentage of the data is greater than $\mu + (1.6)\,\sigma$? Less than $\mu + (1.6)\,\sigma$? What is the probability that a datum picked randomly from the population is greater than $\mu + (1.6)\,\sigma$? Less than $\mu + (1.6)\,\sigma$?
   (c) What percentage of the data is greater than $\mu - (1.4)\,\sigma$ and less than $\mu + (3.2)\,\sigma$? What is the probability that a datum picked randomly from the population will lie between these two numbers?
   (d) What percentage of the population is greater than $\mu + (1.3)\,\sigma$ and less than $\mu + (3.2)\,\sigma$?
   (e) What is the probability that a datum picked randomly from the population will be greater than $\mu - (2.7)\,\sigma$ and at the same time less than $\mu - (1.5)\,\sigma$?

2. In an approximately normal population with mean μ and standard deviation σ, express in terms of μ and σ that datum which cuts off the bottom 10 per cent of the population. The top 10 per cent.

3. An approximately normal population has a mean of 100 and a standard deviation of 1.
   (a) What datum cuts off the top 5 per cent of the population?
   (b) What datum cuts off the bottom 7 per cent of the population?
   (c) If x lies to the left of the mean the same distance as y lies to the right of the mean, and if the region over the interval from x to y amounts to 50 per cent of the total area under the normal curve, what is x?

4. Studies conducted in a large city hospital show that the mean weight of babies born there is 7.5 pounds, and the population of all such babies has a standard deviation of .6 pounds. (Assume that the weights are normally distributed.)
   (a) What percentage of the babies born here weighs more than 8.5 pounds? What is the probability that a baby born here will weigh more than 8.5 pounds? Less than 8.5 pounds?
   (b) What percentage of the babies born here weighs more than 6 pounds? Less than 6 pounds?
   (c) What percentage of the babies born here weighs between 7 and 8 pounds? What is the probability that a baby born here will weigh either more than 8 pounds or less than 7 pounds?
   (d) Within the same hour a baby was born weighing 5.2 pounds, and a baby was born weighing 9.7 pounds. In terms of probabilities, which event was the more unlikely?

5. Intelligence quotients (IQ's) of the total American population are normally distributed with a standard deviation of about 16 and a mean of about 100.
   (a) Find the range of IQ's centered about the mean possessed by 99 per cent of the population. That is, find x, x less than the mean, such that 99 per cent of the area under the normal curve lies between $\mu - x$ and $\mu + x$.

(b) What percentage of the total population have IQ's above 130? Below 70? What is the probability that a randomly selected American would have an IQ greater than 130? Less than 70?

(c) If you assume that there are 50 million Americans, how many Americans might be expected to have IQ's in excess of 150? Less than 50?

6. A national testing company has devised a test with a range of scores from 200 to 900 for which the scores obtained are normally distributed along this range.

    (a) What is the mean score on this test?

    (b) It is known that the score of 650 is the division point between the bottom 70 per cent of the students who take the test and the top 30 per cent. What is the standard deviation of this test?

7. A national test for graduate students has a minimum score of 240 and a maximum score of 800. The test is designed so that the scores obtained are normally distributed over this range.

    (a) What is the mean score obtained on this test? The mode? The median?

    (b) If 68 per cent of the students taking this test get scores not exceeding 565 and the rest get scores of 565 or above, what is the standard deviation?

    (c) What percentage of the students get scores greater than 760?

    (d) A certain university selects its graduate students from among the top 8 per cent of the students as ranked by their scores on this test. What score must a student make on this test in order to be considered for admission by this university?

8. A light bulb is manufactured by a specific technique to have a mean life expectancy of 1000 hours with a standard deviation of 21 hours.

    (a) What is the probability that a bulb manufactured by this process will burn for 1350 hours?

    (b) A bulb with a life expectancy of less than 350 hours is regarded as rejectable. If the manufacturing process turns out more than $\frac{1}{2}$ per cent rejectable bulbs, then the process must be either changed or scrapped. Is the process satisfactory as it stands?

    (c) A company wants to buy bulbs produced by this process if not less than 99 per cent of the bulbs will have a life expectancy between 850 and 1150 hours. Should the company buy the bulbs produced by this technique?

9. There is a mathematics instructor who assigns grades in a very large lecture section according to this procedure. For each examination he computes the mean, m, and the standard deviation, s, of the examination results. Then he assigns grades as follows:

    Scores above $\mu + 3\sigma = $ A+.
    Scores from $\mu + 2\sigma$ to $\mu + 3\sigma = $ A.
    Scores from $\mu + \sigma$ to $\mu + 2\sigma = $ B.
    Scores from $\mu - \sigma$ to $\mu + \sigma = $ C.
    Scores from $\mu - 2\sigma$ to $\mu - \sigma = $ D.
    Scores from $\mu - 3\sigma$ to $\mu - 2\sigma = $ E.
    Scores below $\mu - 3\sigma = $ E−.

(a) What percentage of his students get A's, B's, C's, D's, and E's?

(b) In a class of 200, how many grades of A+ might be expected?

10. A college administers a special placement test to its entering freshmen. The mean score obtained on this test has been 82, and the test scores over recent years have a standard deviation of 6. It is assumed the test results are normally distributed.

(a) The top 10 per cent of the entering class are eligible to enroll in special honors sections of their courses. What is the lowest grade that a student can get and still be eligible to enroll in an honors program of courses?

(b) The bottom 35 per cent of the students who take this test must enroll in a special freshman course. What is the lowest score that a student can get on this test and not have to enroll in this special course?

10.6 A Statistical Test

There are many kinds of statistical tests designed to answer a great variety of questions. One of these that does not involve a great deal of theoretical preparation before it can be used will be discussed in this section. The mathematics behind this example is profound, but the actual workings of the test are not overly complicated. We are going to consider a test that is used to decide whether or not two or more properties are independent. Tests of this kind are used to answer such questions as, "Does cigarette smoking cause cancer?" "Does being overweight contribute to the likelihood of a heart attack?" "Do blondes really have more fun?"

Suppose we had asked 240 college-age students the following two questions:

1. Are you a male?
2. Do you prefer coffee to tea?

Would you expect that the answers to these questions would be related to each other; that is, that they would be dependent? We have constructed a table, called a **contingency table,** that shows the results of having asked this question. See Table 10.9.

We have not supplied the actual survey results in Table 10.9; that is, we have not told you how many males preferred coffee to tea, and so on. The survey was made of 131 males and 109 females; and the survey determined that of this total of 240 students, 140 preferred coffee to tea and 100 preferred tea to coffee. Our goal is to determine whether or not being male has anything to do with coffee-tea preference. To begin our statistical study, which will answer this question, we are going to make a hypothesis that will serve as the basis for our study.

Table 10.9

Do You Prefer Coffee to Tea?	Are You a Male?		
	Yes	No	Totals
Yes	(a)	(b)	140
No	(c)	(d)	100
Totals	131	109	240

The Hypothesis of Independence.
The property of being a male is independent of the property of preferring coffee to tea.

Our analysis is going to tell us whether or not this hypothesis is reasonable.

First, we must fill in the four cells of Table 10.9 with the **expected values** that would be anticipated according to the hypothesis of independence. Since $\frac{100}{240}$ or 42 per cent of the students surveyed preferred tea to coffee and if sex has nothing to do with coffee-tea preference, then 42 per cent of the males should prefer tea, as should 42 per cent of the females. Since 42 per cent of 131 is 55 and 42 per cent of 109 is 46, we enter these expected values in cells (c) and (d) of Table 10.9. The expected values that should go into cells (a) and (b) are 76 and 63 because 76 is $\frac{140}{240}$ or 58 per cent of 131 and 63 is 58 per cent of 109. See Table 10.10.

Now, the expected values are the survey results that we would anticipate if the hypothesis of independence did in fact hold. But our survey did not produce these values. In Table 10.11 we have shown the actual survey results as well as the expected results if independence were accepted. Expected entries have been put in parentheses.

The question that we have to ask is whether or not the actual survey results are different enough from the expected results to warrant our rejecting the independence hypothesis. First, we must determine how much the survey results differ from the expected results. It would not be unreasonable to subtract the expected results from the actual results to determine the amount of difference in each cell.

Table 10.10

Do You Prefer Coffee to Tea?	Are You a Male?		
	Yes	No	Totals
Yes	(76)	(63)	140
No	(55)	(46)	100
Totals	131	109	240

Table 10.11

Do You Prefer Coffee to Tea?	Are You a Male?		
	Yes	**No**	**Totals**
Yes	84 (76)	56 (63)	140
No	47 (55)	53 (46)	100
Totals	131	109	240

But these differences will sometimes be positive and sometimes negative. To turn them all into positive numbers, we shall square them. Next, since a squared difference of say, 4, would be much less significant if the expected result were 100 than if the expected result were 12, we shall divide this squared difference by the expected result. This provides a measure of the difference relative to the expected result. Last, we shall add up these quotients for all the cells.

Here is the procedure for computing a number that is designed to tell us how far from the expected results our actual results have turned out to be.

1. For each cell in the contingency table determine the number

$$\frac{(\text{actual entry} - \text{expected entry})^2}{\text{expected entry}}.$$

2. Sum all the numbers obtained in item 1. This sum will be called the **key number** for the contingency table.

For our example, the key number can be computed by filling in the columns of Table 10.12.

The key number provides a way of obtaining a numerical measure of the way the actual survey results differ from the results that would be expected on the basis of the hypothesis of independence. Note, for example, that if the survey results

Table 10.12 Computing the Key Number

Cell	Actual x	Expected y	Difference $x - y$	Square $(x - y)^2$	Quotient $(x - y)^2/y$
(a)	84	76	8	64	.84
(b)	56	63	−7	49	.78
(c)	47	55	−8	64	1.16
(d)	53	46	7	49	1.07
				Sum =	3.85

turned out to be exactly the expected results (an unlikely event), then the key number would be 0. If the actual results are almost equal to the expected results, then the differences in each cell will be small; the quotients obtained by dividing by the expected values will be smaller still; and their sum will be small. On the other hand, if the actual survey results turn out to be markedly different from the expected results, then the differences in each cell will be large, the quotients will be large, and their sum will be large. Hence the larger the key number, the more likely the two properties being tested are dependent. The nearer the key number is to 0, the more likely the properties being tested will be independent.

The last question that we must answer is, "How near is near?" This question is answered by more advanced statistics than we can consider here. The surprising result is that each contingency table has its own **critical number.** If the key number is greater than the critical number, the hypothesis of independence must be rejected. If the key number is less than the critical number, the hypothesis of independence is accepted. (See Figure 10.19.) The critical number for the contingency table in our example is 3.84 and the key number is 3.85, so we must reject the hypothesis of independence and conclude that the properties of being male and preferring coffee to tea are dependent—being male implies a tendency toward preferring coffee to tea.

Now, an alert student will be asking about the reliability of this conclusion. Remember: Statistical conclusions almost never are absolute and almost always involve a probabilistic hedge of some kind. In fact, there is one here, too, except it was not told to you. When the critical number of 3.84 was presented, we did not explain that the result you would obtain using this critical number would be valid only 95 per cent of the time. Thus the conclusion we obtained using this critical number, that being male and preferring coffee are dependent, is valid at the 95 per cent confidence level. The conclusion may in fact be incorrect, but the chances that it is such are small—not more than 5 per cent.

Now, if a 95 per cent reliable conclusion is not good enough, perhaps we could be better satisfied with 99 per cent reliability. If so, then know that, at the 99 per cent confidence level, the critical number for our contingency table is 6.63. Thus, at the 99 per cent confidence level, we could *not* conclude that being male and preferring coffee are related on the basis of our sample. At this level of confidence the key number is less than the critical number and we would have no reason to reject the hypothesis of independence.

The critical numbers at the 95 and 99 per cent confidence levels will always be 3.84 and 6.63, respectively, if the contingency table is "two-by-two"; that is, if the two properties being compared admit only two possible outcomes, "Yes" and "No." But there are problems in which there may be more than two responses to the question "Do you have property X?" For example, there are at least three distinct responses to the question "Do you agree that ———?" The possible responses are "Yes," "No," and "Not sure." The concluding exercises of this section involve such properties. The critical numbers for each problem are always given both at the 95 per cent and 99 per cent confidence levels.

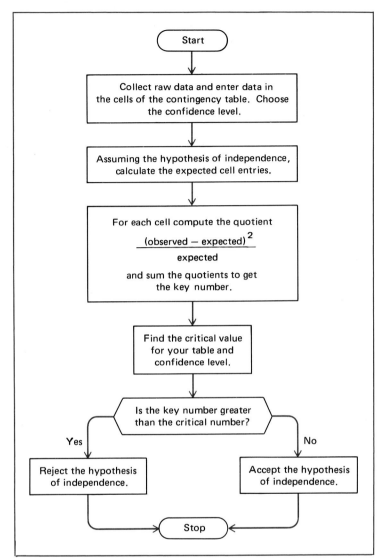

Figure 10.19

This concludes our brief study of statistics. Most schools have elementary courses in statistics; if these ideas interest you at all, inquire of your counsellors whether there might be such a course suitable for you. Statistics is not only interesting for its own sake, but a study of the subject can provide an insight into everyday living that is invaluable.

Exercises 10.6

1. Here are some contingency tables with the entries left out. All you know are the totals in the various columns and rows. If you assume that the two properties in each table are completely independent of one another, what would you expect the entries in these tables to be?

Property B	Property A		
	Yes	No	Totals
Yes			0
No			100
Totals	100	0	100

Property B	Property A		
	Yes	No	Totals
Yes			50
No			50
Totals	60	40	100

Property B	Property A		
	Yes	No	Totals
Yes			140
No			60
Totals	120	80	200

Property B	Property A		
	Yes	No	Totals
Yes			115
No			245
Totals	265	95	360

2. Each test is based upon a *hypothesis of independence*. State the hypothesis that would be the basis for the test applied to each of these pairs of properties.
   (a) The property of being a female and the property of habitually using perfume.
   (b) The property of being married and the property of owning your own home.
   (c) The property of being good at arithmetic computation and the property of being an accountant.
3. In the following contingency table the actual entries have been inserted. Complete the table by adding in the totals and then draw a conclusion as to whether or not the properties A and B are independent or dependent.

Property B	Property A		
	Yes	No	Totals
Yes	30	20	
No	90	60	
Totals			

4. The actual entries are given in the following contingency table and the entries that would be expected on the assumption of the hypothesis of independence are given in parentheses.

Property B	Property A		
	Yes	No	Totals
Yes	35 (40)	45 (40)	80
No	65 (60)	55 (60)	120
Totals	100	100	200

(a) Compute the key number for this contingency table.
(b) If the critical number for this table were 3.84, would the properties be dependent or independent? What would be the case if the critical number were 6.63?

5. A mathematics department polled 240 freshman calculus students on the subject of their intention to take courses in mathematics beyond the calculus and their expected grades in calculus. Here are the results of the poll.

Do You Plan to Take Post Calculus Mathematics Courses?	Do You Expect to Get a "Good" Grade in Calculus?		
	Yes	No	Totals
Yes	110 ( )	35 ( )	115
No	70 ( )	25 ( )	125
Totals	180	60	240

(a) Complete this contingency table by adding the expected entry in each cell.
(b) At the 99 per cent confidence level (critical number — 6.63), do we find a dependence relationship between the students' answers to these questions?

6. A park district is anticipating going to the voters for permission to float a bond issue to improve the city parks. The directors of the district would like to identify from which portion of the voters they will meet the greatest opposition. A suggestion has been made that the greatest opposition will come from the homeowners because they would feel the cost of the bond issue directly and visibly through their property taxes. To test whether or not being a homeowner bears a relationship to being against the bond issue, the directors take a poll. Here, presented in the form of a contingency table, are the results of this poll.

Do You Favor the Bond Issue?	Are You a Homeowner?		
	Yes	No	Totals
Yes	65 ( )	40 ( )	105
No	40 ( )	10 ( )	50
Undecided	25 ( )	20 ( )	45
Totals	130	70	200

(a) Complete this contingency table by adding the expected values in each of the six cells.

(b) Determine the key number for this table.

(c) At the 95 per cent confidence level the critical number for this table is 5.99, whereas at the 99 per cent confidence level the critical number is 9.21. What should the park district directors conclude on the basis of this polling of 200 citizens?

7. The following contingency table is self-explanatory. Complete the table by adding the expected values in each cell and compute the key number for the table. This table has a critical number of 9.49 at the 95 per cent confidence level, and a critical number of 13.28 at the 99 per cent confidence level. What conclusions may be drawn about the dependence of the two properties being tested?

How Many Times in the Past Year Have You Used Mass Transit Facilities?	Do You Favor the Diversion of Motor Fuel Tax Funds to Rapid Transit Authorities?			
	Yes	No	Undecided	Totals
More than 10 times	35 ( )	5 ( )	10 ( )	50
More than 3 times but less than 10 times	35 ( )	15 ( )	20 ( )	70
Less than 3 times	70 ( )	110 ( )	100 ( )	280
Totals	140	130	130	400

Answers to Selected Exercises

Exercises 0.1

1. (a) Turn off, turn on. Some lamps can accept an instruction to change the level of their illumination.
   (b) None.
   (c) All sorts of instructions such as "Print the letter A," "Leave a blank space," and so on.
   (e) A telephone may be given millions of instructions. For example, "Connect yourself with the telephone belonging to Mr. James Perkins in Waukegan, Illinois."
2. Coin machines in general. There are plans to make it possible for telephones to understand such instructions as, "From the account of Mr. X send $20 to the account of the ABC Company." Some tape recorders are made to be self-starting; that is, they automatically begin to record at the first sound signal that reaches them.
3. Coin machines in general discriminate at some level. Dollar bill changers are fairly sophisticated discriminators. Locks are made to discriminate among keys.

Exercises 0.2

1. (a)

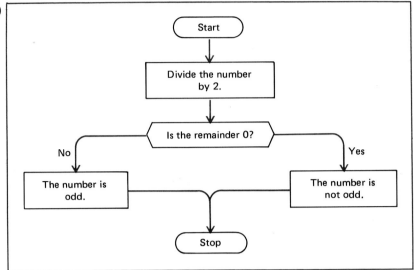

(b)

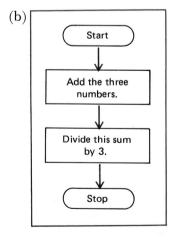

(c)

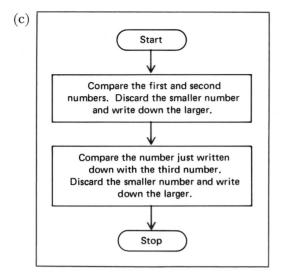

Start

Compare the first and second numbers. Discard the smaller number and write down the larger.

Compare the number just written down with the third number. Discard the smaller number and write down the larger.

Stop

(d)

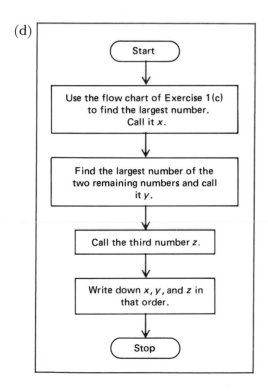

Start

Use the flow chart of Exercise 1(c) to find the largest number. Call it x.

Find the largest number of the two remaining numbers and call it y.

Call the third number z.

Write down $x, y,$ and z in that order.

Stop

2. Since the actual instructions for each operation are only vague, there are many ways to make these flow charts. The following are only examples of the kinds of answers that are possible.

(a)

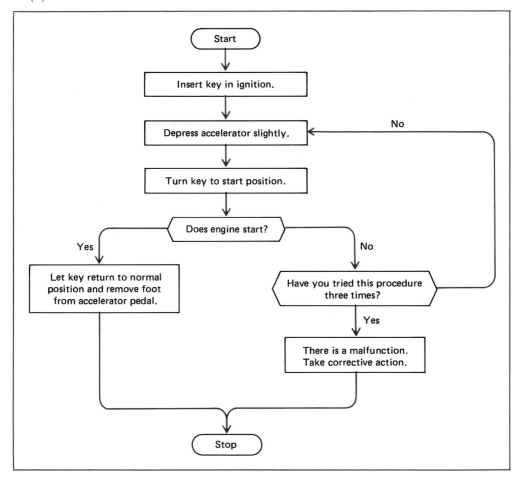

(c)

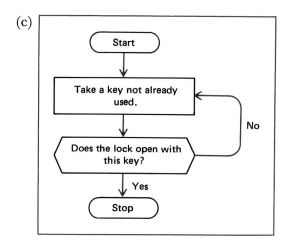

(e)

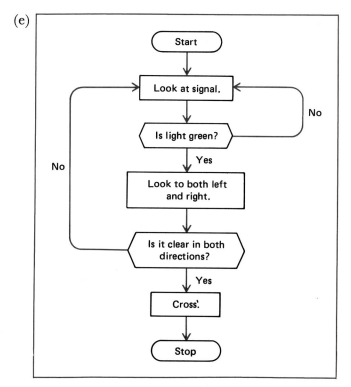

3. No. If a flow chart did have a loop not followed by a decision step, then there would be no way to leave the loop—you would simply run round and round the loop forever. Flow charts *must* terminate at some stage. They can be used only for describing processes that are finite in the sense that they do not go on and on without having some definite predictable ending place.

Exercises 1.1

1. No, the reasoning is inductive, not deductive.
2. John is in the library. You are using *deductive* reasoning.
7. No. Axiom 3 says $10 \cdot 6 = 4$, while axioms 1 and 2 together with axiom 5 combine to prove $10 \cdot 6 = (8 \cdot 2) \cdot (5 \cdot 2) = 44$.
8. (a) The axiom "$2 + 4$ is even" can be deduced from "The sum of two even numbers is even," since we know that 2 and 4 are even.
10. This definition is circular because it tries to define the word *panda* using the word *panda*. So the definition would assume that you know what a panda is before you know what it is.

Exercises 1.3

1. (a) $p \vee q$.
   (b) $\sim p \wedge q$.
   (c) $p \vee \sim q$.
   (d) $\sim(\sim p \wedge q)$.
   (e) $\sim p \wedge q$.
3. (a) $2 = 3 + 7$ or Henry VIII was a king of England. True, since Henry VIII was a king of England.
   (b) It is not true that either Henry VIII was a king of England or $2 = 3 + 7$. False, since Henry VIII was a king of England. (Check with a truth table for $\sim(q \vee p)$.)
4. (a) I do not like school and I do not like to study.
   (b) I like school and do not like to study.
   (c) Either I hate school or I hate to study.

7.

p	q	$\sim p$	$\sim p \vee q$	$p \wedge (\sim p \vee q)$
T	T	F	T	T
T	F	F	F	F
F	T	T	T	F
F	F	T	T	F

Exercises 1.4

2. If $p \to q$ and $q \to p$ are both true, then either both p and q are true or both p and q are false. If $p \to q$ and $q \to p$ have opposite values, then p and q have opposite values. See Table 1.7.

4.

p	q	$\sim p$	$\sim q$	$p \to q$	$\sim q \to \sim p$
T	T	F	F	T	T
T	F	F	T	F	F
F	T	T	F	T	T
F	F	T	T	T	T

The truth tables are the same, so $p \rightarrow q$ and $q \rightarrow \sim p$ are logically equivalent.

5. (a) If $x \neq 4$, then $x + 3 \neq 7$.
   (c) If the ground does not get wet, it will not be raining.
   (e) If a number is an even number, then it is a multiple of two.

7. (a) No, $p \rightarrow (p \wedge q)$ is not a tautology.
   (b) Yes, it is a tautology.
   (c) No, it is not a tautology.

9. (a) If I spend money, I am shopping. If I do not spend money, I am not shopping. If I do not go shopping, I shall not spend money.
   (c) If I save nine stitches, I take one in time. If I do not save nine stitches, I shall not take one in time. If I do not save a stitch in time, I shall not save nine.

11. (a) False. DeMorgan's first law says $\sim[(2 < 3) \wedge (2 > 3)] = \sim(2 < 3) \vee \sim(2 > 3)$. So the true statement is "Either $2 \nless 3$ or $2 \ngtr 3$."
   (c) DeMorgan's second law yields "The solution of the equation $x + 3 = 5$ is neither 6 nor 7."

Exercises 1.5 _____

1. (a) Valid. $\sim(p \vee q) \longleftrightarrow \sim p \wedge \sim q$ (DeMorgan's second law); $(\sim p \wedge \sim q) \rightarrow \sim q$ (law 5).
   (b) Valid. Law 6 $(q \rightarrow (p \rightarrow q))$ and the law of detachment $(q \wedge [q \rightarrow (p \rightarrow q)] \rightarrow (p \rightarrow q).)$
   (c) Valid.
   $$(p \rightarrow q) \wedge \sim q \longleftrightarrow (\sim q \rightarrow \sim p) \wedge \sim q \qquad \text{(law 7)}$$
   and
   $$[(\sim q \rightarrow \sim p) \wedge \sim q] \rightarrow \sim p \text{ (law of detachment).}$$

2. (a) The law of detachment, applied to $2 < 3$ and $(2 < 3) \rightarrow (3 < 4)$, yields $3 < 4$. Another application yields $4 < 5$.
   (b) The law of syllogism tells us "If horses give milk, then people will get mad."
   (c) The law of detachment, applied to $p \rightarrow q$ and p, yields q. The law of syllogism, applied to $q \rightarrow r$ and $r \rightarrow s$, gives $q \rightarrow s$. The law of detachment, applied to q and $q \rightarrow s$, yields s.
   (d) The law of syllogism, applied to $(r \rightarrow s) \rightarrow (p \wedge q)$ and $(p \wedge q) \rightarrow (p \vee q)$, yields $(r \rightarrow s) \rightarrow (p \vee q)$, which together with $r \rightarrow s$ and the law of detachment gives $p \vee q$.

4. (a) Let p be "$2 + 2 = 5$" and q be "$2 = 3$." Conclude "$2 + 2 \neq 5$."
   (c) Let p be "$3 + 5 < 9$" and q be "Horses fly." Conclude "$3 + 5 \nless 9$."

5. (a) Let p be "Cows give milk" and q be "Cows fly." Conclude "Cows fly."
   (c) Let p be "Ethelred was king" and q be "Ethelbert was not king." Conclude "Ethelbert was not king."

6. (a) Let p be "Horses fly" and q be "$x = y$." Modus tollendo ponens gives

"$x = y$." Now let p be "$x \neq 0$" and let q be "$x \neq y$." Modus tollendo tollens gives $x = 0$.

(c) Modus tollendo tollens gives "John does not love Mary." Modus tollendo ponens gives "Horses live in houses." Finally, the law of detachment yields "$x = 0$."

Exercises 1.6

1. (a) There exists a man who works hard and there exists a man who works very hard.
   (c) There exists a man who worked hard and every man sleeps.
   (e) There exists a number that when squared is equal to itself.
2. (a) There exists a number less than zero.
3. (a) Every implication is either true or false.
   (c) Every true implication has a true contrapositive.
7. No. For example, "Every unicorn is a four-legged animal." There are no unicorns.

Exercises 2.1

1. (a) Dividing y by x is an operation. "Being a divisor of" is a relation.
   (b) An operation.
   (c) An operation.
   (d) A relation.
   (e) A relation.
2. The number 5 is related to 25 by the relation "is the square root of" and "is one-fifth of." The number 25 is related to 5 by the relation "is greater than" and "is the square of."
3. "Is 2 less than." The number 5. The number 9. Yes, 2. No, there are no *whole numbers* less than 0.
5. The relations "is congruent to" and "is similar to."
8. The operation performed on x and y yields the product of $\sqrt{x}$ and y. The result of the operation on 25 and 36 is $5 \cdot 36 = 180$. The result of the operation on 36 and 25 is $6 \cdot 25 = 150$. $180 \neq 150$, so the operation is *not commutative*.

Exercises 2.2

1. No. If Paul is the father of John, then John is not the father of Paul, so the relation is not symmetric. It is neither reflexive nor transitive.
2. "Is the sibling of" has the symmetric and transitive properties, whereas "is the brother of" and "is the sister of" have only the transitive property.
4. Yes.

9. The relation "is inscribed inside" or "has smaller area than."
10. Symmetry only.
11. The flow chart must describe a process that eventually terminates after a finite number of steps. If the flow chart of Figure 2.4 were to be applied to a relation defined on an infinite set of objects (such as the set of all whole numbers), then the process described by the chart would never terminate (there would always be unexamined objects); and a final decision about reflexivity would never be made.

Exercises 2.3

1. (a), (b), (d) and (g) are binary operations. The remainder are *not*.
2. The average of 2 and 5 is $\frac{7}{2}$, which is *not* a whole number.
4. (a) x is a diamond.
   (b) x and y have the same rank.
   (c) x and z have the same suit.
   (d) Yes, $5\diamondsuit * 5\heartsuit = 5\diamondsuit$. Yes, $7\diamondsuit * 9\diamondsuit = 9\diamondsuit$.
5. Yes. If p and q are true, then $p \wedge q$, is true. Yes. If p and q are false, then $p \wedge q$ is false as well.
8. (b), (d), and (e) are closed with respect to the addition of whole numbers. (a) and (c) are not since, for (a), $1 + 5 = 6$ (which is not in the given set); and for (c), $1 + 5 = 6$ (which is not odd).
10. $3\diamondsuit * K\clubsuit = K\diamondsuit$; $K\clubsuit * 5\diamondsuit = 5\clubsuit$; all other combinations are already in the set S.

Exercises 2.4

1. (a) Commutative.
   (b) Commutative.
   (c) Associative.
3. (a) $6 \cdot (3 \cdot 2) = (6 \cdot 3) \cdot 2 = (3 \cdot 6) \cdot 2$; associative and commutative laws of multiplication, respectively.
   (b) $6 + (2 + 3) = (6 + 2) + 3 = (2 + 6) + 3$; associative and commutative laws of addition, respectively.
   (c) Commutative and associative laws of addition.
   (d) Commutative, associative, and commutative laws of addition.
5. Yes, both conjunction and disjunction are commutative. Examine truth tables.
7. Let p be "$2 + 2 = 4$," q be "$3 = 5$," and r be "$1 < 7$." Then $p \rightarrow q$ is false, but $q \rightarrow p$ is true; $(p \rightarrow q) \rightarrow r$ is true, but $p \rightarrow (q \rightarrow r)$ is false.
9. $(3\heartsuit * 6\diamondsuit) * J\spadesuit = J\heartsuit$; $3\heartsuit * (6\diamondsuit * J\spadesuit) = 3\heartsuit * J\diamondsuit = J\heartsuit$.
10. This binary operation is commutative. The table is symmetric with respect to the diagonal; that is, the triangle above the diagonal is the exact mirror image of the triangle below the diagonal.

Exercises 2.5

1. $5\diamond * 4\heartsuit = 4\diamond \neq 4\heartsuit$; so $5\diamond$ cannot be an identity for this operation.
3. Yes. Let i be any false statement. Then $p \vee i$ will have the same value that p has.
4. (a), (b), and (c) are incorrect; (d) is correct; (e) is correct if $a \neq b$. (f) is correct if $a \neq (-b)$; and (g) is correct if $a \neq (-b) - 1$.
6. In vertical multiplication we first multiply 34 by 8 and then multiply 34 by 10. So, we actually perform $34(8 + 10)$ and do not multiply 34 by 18.
8. Use truth tables to determine that (b) is true, while (a) is false.
9. (1) left; (2) left; (3) left; (4) associative, addition; (5) transitive.
12. (c), (d), (g), and (h) are false; (a), (b), (e), and (f) are true.

Exercises 2.6

1. $(3 \oplus 5) \oplus (4 \oplus 2) \oplus (1 \oplus 3) \oplus 5 = 2 \oplus 0 \oplus 4 \oplus 5 = 2 \oplus 3 = 5.$
3. (a) $x \ominus 2 = 4$ means that $x = 4 \oplus 2$. So $x = 0$.
   (b) $3 \ominus (x \oplus x) = 1$; $3 \ominus 1 = x \oplus x$; $2 = x \oplus x$; so $x = 1$ or $x = 4$.
   (c) $x = 3$.
6. 0 is the additive inverse of 0; 5 is the additive inverse of 1; 4 of 2; 3 of 3; 2 of 4; 5 of 1.
8. If we used this definition, then $2 < 3$ (since $2 \oplus 1 = 3$) and $3 > 2$ (since $3 \oplus 5 = 2$).
11. (a) No solution. (Examine the diagonal.)
    (b) $x = 1$ or $x = 3$.
    (c) $x = 0$ or $x = 2$.
    (d) $x = 3$.
    (e) $x = 1$.
    (f) $x = 2$.

Exercises 2.7

1. $(4 \otimes 4) \otimes (5 \otimes 2) \otimes 5 = (4 \otimes 4) \otimes 5 = 4 \otimes 5 = 2.$
2. Use the (completed) multiplication Table 2.3.
   (a) $x = 1$ or $x = 5$.
   (b) $x = 2$ or $x = 4$.
   (c) $(2 \otimes x) \ominus 3 = 0$; $(2 \otimes x) = 3$; there is no solution, since 2 is even and 3 is odd.
   (d) $x = 1$ or $x = 4$.
   (e) No solution.
   (f) Checking each of the six possible values for x in this equation produces solutions $x = 1$, $x = 2$, $x = 4$, or $x = 5$.

5. $(3 \otimes 4) \otimes 5 = 0 \otimes 5 = 0$;　$3 \otimes (4 \otimes 5) = 3 \otimes 2 = 0$;　so　$(3 \otimes 4) \otimes 5 = 3 \otimes (4 \otimes 5)$. The operation 6-multiplication is also commutative. 1 is an identity.

6. No. There is no y such that $2 \otimes y = 1$.

9. Yes. The inverse of 1 is 1; the inverse of 3 is 2; and the inverse of 4 is 4.

Exercises 3.2

1. (a) $3 < 5$, since $3 + 2 = 5$.
   (b) $5 > 3$, since $3 < 5$.
   (c) $0 < 5$, since $0 + 5 = 5$.
   (d) $5 > 0$, since $0 < 5$.

2. x is greater than or equal to y, $x \geq y$, if either $x > y$ or $x = y$.

5. (Theorem A) If x, y, and z are whole numbers and if $x > y$, then $x + z > y + z$.
   (Theorem B) If x, y, and z are whole numbers, if $x \geq y$, and if $z \neq 0$, then $xz \geq yz$.

7. If z were equal to 0 in Theorem B, then multiplying the true statement $3 < 5$ by 0 on both sides would result in the false statement $0 < 0$.

9. (1) Hypothesis. (2) From (1), using the definition of *is less than*. (3) Substitution of equals for equals in (2). (4) Associative law of addition, applied to (3). (5) d and e are nonzero whole numbers so that $d + e$ is also. (6) From (4) and (5), using the definition of *is less than*.

Exercises 3.3

1. $3 \div 0$ is meaningless, for if $3 \div 0 = q$, q is a whole number, then it would follow that $3 = 0 \cdot q = 0$. But $3 \neq 0$. Also, $3 \div 2$ is meaningless, for if $3 \div 2 = q$, q a whole number, then it would follow that $3 = 2 \cdot q$. But 3 is odd and $2q$ is even. $2 - 3$ is meaningless, since there is no whole number x such that $2 = 3 + x$. $0 \div 0$ is meaningless for, if it were meaningful, then because $0 = 0 \cdot q$ is true for every whole number q, $0 \div 0$ would have to be equal to every whole number. Thus we could not find a *unique* whole number that was the quotient. $0 \div 3$ is meaningful; it is equal to 0, since $0 = 3 \cdot 0$.

3. (b) and (d) are true. (a), (c), (e), and (f) are false. Multiplication is ambidextrous in the sense that we have a left and right distributive law for both addition and subtraction. That is, (b) is true, whereas (a) is false.

4. (a) $x + 6 = 13$; $(x + 6) - 6 = 13 - 6$; $x = 7$ by Theorem C.
   (c) $3x - 5 = 10$; $(3x - 5) + 5 = 10 + 5$; $3x = 15$ by Theorem D; $(\frac{1}{3})(3x) = (\frac{1}{3})(15)$; $x = 5$ by Theorem E.

5. (a) $x + 2 < 7$; $(x + 2) - 2 < 7 - 2$ by the analog of Theorem A; $x < 5$ by Theorem C.

6. Commutative property of multiplication; definition of division.

Exercises 3.4

1. (a) $(x + 2)(x - 2) = 0$. By Theorem H either $x + 2 = 0$ or $x - 2 = 0$. But there is no whole number x such that $x + 2 = 0$, so there is no solution to $x + 2 = 0$. The solution of $x - 2 = 0$ is $x = 2$. Hence 2 is the only solution of the equation.

2. Statement (6) does not follow from Statement (5), since if $a = b$, then $a - b = 0$ and we cannot cancel 0 like this.

Exercises 3.5

1. (a) $2^2 \cdot 3^3$.
   (c) $2^2 \cdot 3^2 \cdot 5^3 \cdot 7^3$.
2. (a) $2^6 \cdot 3^5 \cdot 5^4$.
   (b) $1 \cdot 1 = 1$.
   (c) $2^6 \div 2^3 = 2^3$.
3. $E = mcc$. $E = (mc)^2 = mcmc$ is what the equation does not mean.
5. 2 parents; $4 = 2^2$ grandparents; $8 = 2^3$ great grandparents; $16 = 2^4$ great-great grandparents; 32 great-great-great grandparents. $2 + 2^2 + 2^3 + 2^4 + 2^5 = 62$ ancestors, since in 200 years there are 6 generations.
7. 1 googol $= 10^{100}$. 1 googolplex $= 1$ followed by 10^{100} zeros. Therefore, a googolplex is $10^{googol} = 10^{(10^{100})}$.

Exercises 3.6

1. Divisors of 12 are 1, 2, 3, 4, 6, and 12. Divisors of 145 are 1, 5, 29, and 145. Divisors of 144 are 1, 2, 3, 4, 6, 8, 9, 12, 18, 24, 36, 48, 72, and 144.
4. $38 = 7 + 31$, $40 = 11 + 29$, $42 = 13 + 29, \ldots,$ $148 = 11 + 137$, $150 = 13 + 137$, $638 = 19 + 619$, $640 = 23 + 617$, and $642 = 41 + 601$.
6. (a) 1, 2, 4, 8, and 16 are divisors of 16.
   (b) 16 is a divisor of 16, 32, 48, 64, 80, ...
   (c) Yes; each whole number is a divisor of itself.
   (d) Yes; if x is a divisor of y and y is a divisor of z, then x is a divisor of z.
   (e) 3 is a divisor of 6, but 6 is not a divisor of 3.
8. $51! + 2 = 2[(51! \div 2) + 1]$; $51! + 3 = 3[(51! \div 3) + 1]$;

$$51! + 4 = 4[(51! \div 4) + 1], \ldots, 51! + 51 = 51[(51! \div 51) + 1].$$

Since there are 50 numbers between 2 and 51 inclusive, there are 50 numbers in this list. To compile 100 successive composite numbers, examine $101! + 2$, $101! + 3, \ldots, 101! + 101$.

10. (a) Let $n = 3$. $2^3 + 1 = 9$, which is not prime.
    (b) Let $n = 4$. Then, $2^4 - 1 = 15$, which is not prime.
    (c) Let $n = 11$. Then, $121 - 11 + 11 = 121$, which is not prime.
    (d) Let $n = 41$. Then, $(41)^2 - 41 + 41 = (41)^2$, which is not prime.

Exercises 3.7

2. 5^0 is 1, which is not a prime.
3. $58 : 1, 5, 5^2, \ldots, 5^{57}$.
5. If every prime in the prime factorization of a whole number appears with an even exponent, then that number is a perfect square. Its square root is the number you get by halving each of these even exponents.
6. A whole number is square-free if each prime appearing in the number's prime factorization has the exponent 1. 2, 3, 5, $6 = 2^1 \cdot 3^1$, 7, 10, 11, 13, 14, 15, 17, 19, 21, 22, 23, 26, 29, $30 = 2^1 \cdot 3^1 \cdot 5^1$.
7. (a) 2, 4, 8.
   (c) 2, 4.
   (e) 2, 4, 5, 10.
8. (a) Divisible by 2, 3, 4, 5, 8, 9, and 10. $1080 = 10 \cdot 108 = 2 \cdot 5 \cdot 4 \cdot 27 = 2^3 \cdot 3^3 \cdot 5$.
   (c) Divisible by 2, 3, and 9. $2754 = 2 \cdot 1377 = 2 \cdot 9 \cdot 153 = 2 \cdot 9 \cdot 9 \cdot 17 = 2 \cdot 3^4 \cdot 17$.

Exercises 3.8

1. (a) $12 = 2^2 \cdot 3$ and $30 = 2 \cdot 3 \cdot 5$. GCD is $2 \cdot 3 = 6$. LCM $= 2^2 \cdot 3 \cdot 5 = 60$.
   (c) $198 = 2 \cdot 3^2 \cdot 11$ and $144 = 2^4 \cdot 3^2$. GCD $= 2 \cdot 3^2 = 18$.

   $$\text{LCM} = 2^4 \cdot 3^2 \cdot 11 = 1584.$$

3. Prove it, using Exercise 2.
4. 101, 145, 169 are relatively prime to 144.
5. (a) No; let $a = d$ and $b = c$. Then $ac = bd$, so that they are not relatively prime.
   (b) No; 2 and 1 are relatively prime; 3 and 4 are relatively prime; but $2 + 3 = 1 + 4$ so that $a + c$ and $b + d$ are not relatively prime.
   (c) Yes; since a prime divides a^2 if and only if it divides a itself.
   (d) Yes; if a prime divides a, it does not divide b, so it cannot divide $a + b$. Also, if a prime were to divide $a + b$ and a, then it would have to divide b, contradicting the fact that a and b are relatively prime.
6. (a) Yes; * is a binary operation, since prime factorizations are unique.
   (b) $\text{GCD}(x, y) = \text{GCD}(y, x)$, so * is commutative. $\text{GCD}(\text{GCD}(x, y), z) = \text{GCD}(x, \text{GCD}(y, z))$, so * is associative. (Try this last equation with a few triples of numbers to get the feel of what it says.)

(c) No; if $x * i = x$, then x divides i. There is no whole number i which is divisible by *every* other (nonzero) whole number.

(d) No; $15 * 25 = 5 = 30 * 25$, but $15 \neq 30$.

Exercises 3.9

1. (a, b, c) a Pythagorean triple means $a^2 + b^2 = c^2$. Then $(2a)^2 + (2b)^2 = 4a^2 + 4b^2 = 4(a^2 + b^2) = 4c^2 = (2c)^2$. Therefore, $(2a, 2b, 2c)$ is also a Pythagorean triple. Also, $(na)^2 + (nb)^2 = n^2a^2 + n^2b^2 = n^2(a^2 + b^2) = n^2c^2 = (nc)^2$, so that (na, nb, nc) is a Pythagorean triple. There are infinitely many Pythagorean triples, since $(3, 4, 5)$ is one and we can obtain another for each whole number n.

3. If a, b, and c are odd, then a^2, b^2, and c^2 are odd. Therefore, $a^2 + b^2$ is even and cannot equal the odd number c^2. Hence not all three numbers of a Pythagorean triple are odd. Similarly, we cannot have two numbers even. However, all three may be even [for example, $(6, 8, 10)$].

Exercises 4.1

2. (a) A blue shirt.
   (b) A white shirt.
   (c) A black table.
   (d) An elephant's tusk.

3. Set $S = \{$Jim, Bob, Marge, Pat$\}$. If $x =$ Jim, then $x \in S$. If $y =$ Harry, then $y \notin S$.

5. (a) $3 \in S$.
   (c) $13 \notin S$.
   (e) $12 \in S$.

6. (a) The set of whole numbers less than 0.

8. (a) $\{14, 15, 16, 17, 18, 19, 20, \ldots\}$.
   (b) $\{2, 1, 0\}$.

9. (a) $\{0, 2, 4, 6, 8\}$
   (b) $\{0, 1, 2, 3, 4, 5, 6, 7, 8, 9\}$.

10. (a) $\{3, 4\}$.
    (b) $\{0, 1, 2\}$.

Exercises 4.2

2. The given expression, which names a whole number not namable in less than 22 syllables, does it in only 19 syllables.

3. The barber was a female, since if the barber were a man, he could not shave himself if he shaved himself—a paradox.

4. If "heterological" were autological, it would describe itself, but "heterological"

means it does not describe itself—a contradiction. If "heterological" were heterological, it would not describe itself, but that's what "heterological" means, so it would be autological as well—a contradiction. Hence "heterological" (and "autological") is neither heterological nor autological.

Exercises 4.3

1. (a), (b), and (c) are proper subsets of $\{1, 2, 3\}$; (c) and (g) are disjoint from $\{1, 2, 3\}$; (e) is equal to $\{1, 2, 3\}$; (a), (b), (c), and (d) are subsets of $\{1, 2, 3\}$.
2. A set containing n elements has 2^n subsets.
3. One—S itself.
4. (a)

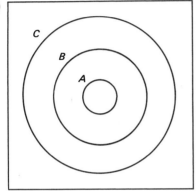

5. (a)

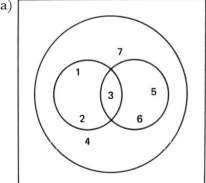

7. Yes. Let $A = \varnothing$ and $B = \{1, 2, 3\}$. Then $A \subset B$, but $A \cap B = \varnothing$.
9. No, since $\{1\} \cap \{1, 2\}$ is $\{1\}$, not $\varnothing$. $\{1\}$, $\{2\}$, $\{3\}$, $\{4\}$, $\{5\}$, and $\{6\}$ are pairwise disjoint sets.
10. Symmetric, not reflexive, not transitive. See the figure at the top of p. 330.
12. (a) and (c); (b), (d), and (e).

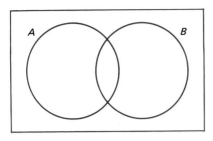

Exercises 4.4

1. (a) $A \cap (B \cup C) = \{1, 2, 3\} \cap \{0, 2, 3, 4, 5, 6\} = \{2, 3\}$.
   (c) $A \cup (B \cup C) = \{1, 2, 3\} \cup \{0, 2, 3, 4, 5, 6\} = \{0, 1, 2, 3, 4, 5, 6\}$.
3. $A \cup B = B$; $A \cup C = C$; $B \cup C = C$.
4. $A \cup B = B$ means A is a subset of B. But if $A \cap B = \varnothing$, then A is $\varnothing$.
7. Let $A = \{1, 2\}$, $B = \{2, 3\}$ and $C = \{1, 2, 3\}$. Then

$$A \cup C = \{1, 2, 3\} = B \cup C.$$

Since $A \neq B$, then the union operation does not have the cancellation property.
9. The 13 people of Muck belong to Klutz and Smorg, and so are counted twice. The total population of Klutz and Smorg is $7891 + 9047 - 13 = 16{,}925$.
11. (a) $\{1, 2, 3\} - \{2, 3\} = \{1\}$; $\{1, 2, 3\} - \varnothing = \{1, 2, 3\}$;

$$\{1, 2, 3\} - \{5, 6, 7\} = \{1, 2, 3\}.$$

   (b) $A \cap B = \varnothing$.
   (d) No; for example, $\{1, 2\} - \{1\} = \{2\}$, but $\{1\} - \{1, 2\} = \varnothing$.
13. Let C = students in chemistry, B = students in biology, and P = students in physics. Then $C \cap B \cap P = 6$, $C \cap B = 20$, $C \cap P = 38$ and $B \cap P = 29$. So we see that 8 students take only chemistry, 32 take only biology, and 9 take only physics. The total number of students is

$$8 + 14 + 6 = 23 + 9 + 32 = 124.$$

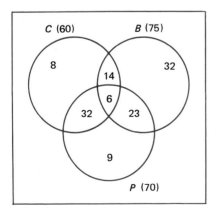

14. Yes, $A = \emptyset$ is an annihilator for intersection. The universal set is an annihilator for union.

Exercises 4.5

1. (a) $a \longleftrightarrow 1, b \longleftrightarrow 2; a \longleftrightarrow 2, b \longleftrightarrow 1$.
   (c) There are none.
3. The whole numbers that are multiples of 3 form an infinite set, since it is in one-to-one correspondence with a proper subset, for example, the multiples of 6.
5. Two students could have exactly the same given name, but if everyone has a different number-name, there will be a one-to-one correspondence between the set of students and the set of number-names of students.
6. (a) No, equivalent sets are not equal. But equal sets are equivalent.
   (b) All three properties.

Exercises 4.6

1. $\aleph_0$ since $\{1, 10, 100, 1000, \ldots\}$ is in one-to-one correspondence with the set of whole numbers, one such correspondence being that which pairs a whole number n with the power 10^n.
2. 6, since there are 6 different one-to-one correspondences between the set $\{a, b, c\}$ and the set $\{1, 2, 3\}$; $(4)(3)(2)(1) = 24$; $(n)(n-1)(n-2)(n-3) \ldots (3)(2)(1)$.
3. (a) 1 as a cardinal.
   (b) 6 as an ordinal.
4. Let A be the set of whole numbers that are multiples of 12. Let B be the set of whole numbers that are multiples of 6. Let C be the set of whole numbers that are multiples of 3. Then $A \subset B \subset C$, and the cardinal of each is $\aleph_0$. All sets must be infinite.
6. Do the same thing as was done in Example 6, using the square in place of the larger circle. This method works for many geometric shapes.
7. S has cardinal number 2; $P(S)$ has cardinal number $2^2 = 4$; $P(P(S))$ has cardinal number $2^4 = 16$; and $P(P(P(S)))$ has cardinal number $2^{16} = 65,536$.

Exercises 4.7

1. The manager must tell him that there is no room for him. Since $1000 + 1 \neq 1000$, Mr. Beaumont cannot be given a room in this hotel. But in a hotel with $\aleph_0$ rooms, one more person can always be added to the register.
2. Let $A = \{1, 2\}$ and $B = \{3, 4, 5, 6, 7\}$. Then the cardinality of $A \cup B = 7$, which is the sum of 2 and 5, since $A \cap B = \emptyset$. Let $C = \{0, 2, 4, 6, 8, 10, \ldots\}$ and $D = \{1, 3, 5, 7, 9, \ldots\}$. Then the cardinality of C is $\aleph_0$, as is the cardinality

of D. The set $C \cup D$ is the set of whole numbers. So $\aleph_0 + \aleph_0 = \aleph_0$, since $C \cap D = \varnothing$.

3. Let $A = \{1, 2, 3, 4, 5, \ldots\}$ and $B = \{1, 2, 3\}$. Then $A - B = \{4, 5, 6, \ldots\}$. So $\aleph_0 - 3 = \aleph_0$.

4. All he has to do is move the occupants of room 1 to room 11, of room 2 to room 12, etc. Then put the occupants of the 10 rooms in Wing A into rooms 1 through 10.

Exercises 5.1

1. (a) A knot; for example, a bowline.
   (b) A highway between two towns.
   (c) The path traveled by the average taxi from the time it leaves the garage in the morning until it returns that evening.
   (d) The edge of a tire.
3. If you examine the edges of a square box, you will be able to find two that lie on what are called "skew" lines; that is, lines that neither intersect nor are parallel.
4. (b) True.
   (c) True.
   (d) False.
6. The earth and the other planets trace out simple closed curves in space as they orbit, as do their respective moons. The tip of the second hand on a clock traces out a simple closed curve in 1 minute.
8. (a) No.
   (b) Yes.
   (c) No.
9. (a) Yes, two tangent circles.
   (b) No, the curves would have to meet in order to form one curve, and then it would not be simple.
11. In 5.6 the triangular and the circular regions are convex. In 5.7, the sphere and the cylinder are convex, while the others are not.

Exercises 5.2

1. Walking across a room is an example of translation; rolling over is an example of reflection; spinning a top is an example of rotation.
2. (a) Translation.
   (b) Rotation, followed by translation.
   (c) Rotation and translation.
   (d) Rotation, reflection, and translation.
   (e) Rotation and translation.
   (f) Two reflections.

4. Straightening a picture frame on a wall; putting a cup in a saucer.
6. Either a rotation about a vertex or a reflection about a line intersecting the triangle at one vertex only.
7. It must have been a reflection about one side.
8. It must have been a rotation of $360°$. That is, it left all the points of the triangle fixed.

Exercises 5.3

1. Cooking a hamburger patty; squeezing a rubber ball; stretching an elastic band.
2. (a) Shrink the left-hand circle and translate.
   (b) Shrink the left-hand line; rotate and translate.
3. It enlarges the figure. It shrinks the figure. It does not change the figure.
6. The ratio must be 1, and all points must be left unmoved by the homothety.
8. Yes.
9. If the area of the original square of side ℓ is ℓ^2, then the area of the square that results from the homothety is $r^2\ell^2$.

Exercises 5.4

1. All are congruence invariants. For example, in (a), if ℓ_1 and ℓ_2 are parallel, and if this configuration is translated in space, the resulting lines are still parallel. All are similarity invariants except (b) and (j).
2. Yes; yes.
3. No. The property of being a square of area 1 cubic inch is a congruence invariant but is not a similarity invariant. Yes. Any property preserved by translations, rotations, reflections, and homotheties is certainly preserved by the first three motions alone.
4. A congruence geometer would see three different circles and three different triangles. A similarity geometer would see only one circle, but would see the same three different triangles that the congruence geometer sees, since no two of these three different triangles are similar.

Exercises 5.5

1. Melting an ice cube; knitting yarn into a scarf; stretching pizza dough into a pan.
3. (a), (b), and (g); (d), (e), and (f); (c) and (h).
6. (i) is a simple closed curve; (a), (c), and (f) are simply connected; (a), (b), (d), (e), (f), (k), and (l) contain circular regions; no figure is a finite set of points; (a), (c), (d), (e), (f), (g), (i), (j), and (l) are arcwise connected; (h) has none of the properties.

7. {A, R}; {B}; {C, G, I, J, L, M, N, S, U, V, W, Z}; {D, O}; {E, F, T, Y}; {H}; {K, X}; {P, Q}.

8. Only (d) and (e) are topological invariants. This illustrates that most of the properties of concern in Euclidean geometry do not even exist in topology.

9. No. Yes. No (a "kidney-shaped region" is simply connected but not convex). Yes. No.

12. Yes. Yes. No, the interior of the simple closed curve would be a hole. Yes.

13. No.

15. (a) Idaho, West Virginia, Minnesota, and Oklahoma are arcwise connected. Hawaii, Michigan, and all coastal states with islands (such as Florida and New York) are not arcwise connected.

   (b) Yes, those that are not arcwise connected.

Exercises 5.6 ──────────────────

1. (a) This is a Moebius strip with a hole cut in it. It has one side and two edges—the edge of the strip and the edge of the hole.

   (b) One side; 2 edges.

3. Draw 2 loop cuts like the second cut in Figure 5.32. The loop number is not 1, since there is a way to make two loop cuts that results in only one piece.

4. The loop number is 0, since any loop cut results in two pieces. (Remember that a square is topologically equivalent to a disk.)

5. It is the same figure as that which results from cutting a Moebius strip down the middle. It has two sides and two edges. You can make the figure by taking two half-turns in the strip before taping.

6. Two sides and two edges. Cutting this figure down the middle results in two of the same figures, intertwined.

7. The figure has one side and one edge. Cutting this figure down the middle results in a one-piece figure that knots around itself. It has one side and one edge. Its loop number is 1.

8. This cut results in two figures hooked together. One is like the original Moebius strip. The other is a strip with two half-turns and is the one-third cut off the "side" of a Moebius strip. (See Exercise 5.6.2.)

Exercises 6.1 ──────────────────

1. $-18, 123, -67, 0, 111$.

2. $(-5) + (-2)$ must equal -7, since when 7 is added to $(-5) + (-2)$, the result is 0 and -7 is the only number with this property: $[(-5) + (-2)] + (7) = (-5) + (-2) + (2) + (5) = (-5) + [(-2) + 2] + (5) = (-5) + 0 + 5 = (-5) + 5 = 0$.

4. $(-111) + 63 = [-48 + (-63)] + 63 = -48 + [(-63) + 63] = -48 + 0 = -48$.

5. $(-6)(0) = (-6)(0) + 0 = (-6)(0) + (6)(0) = (-6 + 6)(0) = (0)(0) = 0.$

7. (3) Adding the same number, $-y$, to both sides of (2). (4) Commutivity of addition. (5) Associativity of addition. (6) y and $-y$ are additive inverses. (7) 0 is the additive identity. (8) This step follows from (1) and (7) by using both the reflexivity and transitivity of equality.

9. (a) $x = 6 - (-4) = 6 + (-(-4)) = 6 + (4) = 10.$
   (b) $2x = (-4) - (-6) = -4 + (-(-6)) = -4 + 6 = 2.$ Hence $x = 1.$

11. (a) $-3 < -2$, since $(-3) + 1 = -2.$

12. (a) If $z > 0$ and $x < y$, then $xz < yz.$

15. (a) 66, 45, 0, 1, 23, 14.
   (b) $|-3| = 3$, so $3|x| = 6$ is the equation to be solved. So $|x| = 2$, which means that x is either 2 or -2.

Exercises 6.2

1. (a) Yes. α is the *-identity; α is its own *-inverse; β is its own *-inverse.
   (b) There are a total of eight equations that must be verified. These include the following three: $(\alpha * \alpha) * \beta = \alpha * (\alpha * \beta)$, $(\alpha * \beta) * \alpha = \alpha * (\beta * \alpha)$, and $(\beta * \beta) * \alpha = \beta * (\beta * \alpha)$. Of these three, the first is verified by independently computing each side: $(\alpha * \alpha) * \beta = \alpha * \beta = \beta$ and $\alpha * (\alpha * \beta) = \alpha * \beta = \beta$ and so the two sides are equal. Since the set together with the binary operation * satisfies 1A, 1B, 1C, and 1D, it forms a commutative group.

3. None of the tables defines a group. The third is not a group because the operation the table defines is not commutative and Exercise 2 states that the operation will have to be commutative. The other three tables do not define groups, because none of these three has an identity element.

4. (a) $(d * b) * c = e * c = d = d * a = d * (b * c)$;
   $(a * f) * b = f * b = d = a * d = a * (f * b).$
   (b) a is the *-identity. The *-inverses of a, b, c, d, e, and f are a, c, b, d, e, and f, respectively.
   (c) $d * c = f$, but $c * d = e$, so $d * c \neq c * d$ and the group is not commutative.

6. If $c * d = c$ rather than e, then we would have $c * d = e * d$, since $e * d = c$. Then if the resulting structure were a group, Exercise 5 would say that since $c * d = e * d$, we must have $c = e$. But $c \neq e$, so the resulting structure cannot be a group.

Exercises 6.3

2. $\frac{2}{3}, \frac{-2}{3}, \frac{11}{8}, \frac{3^2 \cdot 7^2 \cdot 11}{2 \; 5 \; 13}.$

3. (a) Add $\frac{3}{4}$ to both sides to obtain $2x = \frac{5}{7} + \frac{3}{4} = \frac{41}{28}$. Then multiply both sides by $\frac{1}{2}$ to obtain $x = \frac{41}{56}.$
   (b) Write $x = \frac{2}{3} \div (\frac{1}{2} + \frac{2}{3}) = \frac{2}{3} \div (\frac{7}{6}) = \frac{2}{3} \cdot \frac{6}{7} = \frac{12}{21} = \frac{4}{7}.$

5. -7, $\frac{4}{-5}$, $\frac{-2}{3}$, $\frac{-1}{9}$, $\frac{-1}{16}$, 0, $\frac{1}{5}$, $\frac{1}{2}$, $\frac{2}{3}$, $\frac{4}{5}$, 12.

6. There is no positive smallest rational number, for if a/b were such a number, then $a/2b$ is still a rational number and is positive, but is smaller than a/b. This would be a contradiction. The number 1 is the smallest positive integer. There is no largest negative rational number, but -1 is the largest negative integer.

8. (a) $a/b + c/d = ad/bd + bc/bd = (ad + bc)/bd$.

9. (a) $\frac{2}{7} = \frac{1}{7} + \frac{1}{7} = \frac{1}{7} + \frac{1}{8} + \frac{1}{56}$, since $\frac{1}{7} = \frac{1}{8} + \frac{1}{56}$.

   (b) $\frac{3}{7} = \frac{1}{7} + \frac{2}{7}$. Now use the result of part (a).

Exercises 6.4

2. Property 2D fails to hold, since 2 has no multiplicative inverse. (The multiplicative identity is 1.)

3. The group of integers does possess the order property. The whole numbers do not, for properties 4B and 4C cannot hold, since there are no negative whole numbers.

4. In general, when we say that either statement A is true or statement B is true, we allow for the possibility that both statements are true. For example, the disjunction $p \vee q$ is true if p and q are both true. In 4C, we mean to say that *only* one of the statements "$0 < x$," "$x < 0$," and "$x = 0$" is true. Thus the "either-or" terminology is misleading.

Exercises 6.5

1. (a) $x = .2222\cdots$; so $10x = 2.2222\cdots$, whence $9x = 2.0000\cdots = 2$. Thus $x = \frac{2}{9}$.

   (b) Set $x = 4.323232\cdots$. Then $100x = 432.323232\cdots$ and $99x = 428.0000\cdots$ $= 428$. Hence $x = \frac{428}{99}$.

2. (a) $\frac{4}{10} = \frac{2}{5}$. Here $b = 2^0 \cdot 5^1$ and so $\frac{4}{10}$ will have a terminating decimal numeral. In fact, $\frac{4}{10} = .4$.

   (b) $\frac{4}{9} = 2^2/3^2$. Here $b = 3^2$, so $\frac{4}{9}$ has a repeating decimal numeral. $\frac{4}{9} = .4444\cdots$.

5. The decimal numeral of $\frac{1}{99}$ is $.010101\cdots$, while that of $\frac{33}{99}$ is $.3333\cdots$. This does not contradict the theorem stated in Exercise 4, since 33 and 99 are not relatively prime; their GCD is 33. Another such pair is $\frac{1}{75}$ and $\frac{15}{75}$.

Exercises 6.6

2. If one leg of a triangle has length 1 and the hypotenuse has length 2, then the other leg has length $\sqrt{3}$. Lay this triangle off on the number line so that the second leg of the triangle lies along the number line with its left end at the origin. The other end lies at the point corresponding to $\sqrt{3}$.

3. If a right triangle had legs of length a, its hypotenuse would have length $a\sqrt{2}$. If a is an integer, then $a\sqrt{2}$ is irrational. Hence not all three sides of an isosceles right triangle can be of integral length.
6. (a) Irrational.
   (b) Rational.
   (c) Rational. It is equal to 1.
   (d) Rational.
8. π and $1 - \pi$ are both irrational, yet their sum is 1, which is rational. So the sum of two irrational numbers is not necessarily irrational. Neither does the product of two irrational numbers have to be irrational. For example, $(\sqrt{2})(\sqrt{2}) = 2$.

Exercises 7.1

1. (a) $[(0-4)^2 + (3-(-3))^2]^{1/2} = (16+36)^{1/2} = 2(13)^{1/2}$.
   (c) $[(c-a)^2 + (d-b)^2]^{1/2}$.
3. The fourth vertex must have an x-coordinate of -1 and a y-coordinate of 1. So the point is $(-1, 1)$. The x-coordinate of the center must be the same as the x-coordinate of the point midway between $(-1, -3)$ and $(3, -3)$, and so is 1. The y-coordinate of the center is -1.
5. Since the radius is the distance from the center to any point on the circle, we need only determine the distance from $(3, 1)$ to $(7, -2)$. This distance is 5.
7. We must show that two sides have the same length. The lengths of the sides are $\sqrt{50}$, $\sqrt{37}$, and $\sqrt{37}$. Hence it is an isosceles triangle.

Exercises 7.2

1. (a) The difference $x - y$ is always 2, so the equation is $x - y = 2$.
   (b) $2x + y = 2$.
2. (a) $4x + 2y = 3$.
5. It is parallel to the y-axis. It is parallel to the x-axis.
6. Both the x- and y-intercepts are $(0, 0)$ so that the line goes through the origin.

Exercises 7.3

1. (a) Slope $= (2 - (-2))/(2 - 4) = -2$.
   (b) Slope $= (2 - 0)/[(-1) - (-4)] = \frac{2}{3}$.
3. The line is parallel to the y-axis. The line is parallel to the x-axis. The line rises. The line falls.
5. It is either 1 or -1.
8. We know that the altitude is perpendicular to $\overline{BC}$. The slope of $\overline{BC}$ is $\frac{10}{9}$. So the slope of the altitude must be $\frac{-9}{10}$. The equation has the form

$y = (\frac{-9}{10})x + b$. Since the point $(-4, 2)$ is on the line, its coordinates satisfy the equation. Thus $2 = (\frac{-9}{10})(-4) + b$ so that $b = \frac{-8}{5}$. Thus the equation is, in best form, $9x - 10y = -16$.

9. Since the growth is linear, its graph is a straight line. The slope of the line is $(24 - 13)/(67 - 65) = \frac{11}{2}$. On the other hand, the slope from $(67, 24)$ to $(71, y)$ is $(y - 24)/(71 - 67) = (y - 24)/4$. Then we get $(y - 24)/4 = \frac{11}{2}$ so that $y = 46$. The population in 1971 should have been 46.

11. No. The growth would be greater the longer the money were left on deposit. If the interest were withdrawn as soon as it were earned, however, this would not be true.

12. At a constant speed the amount of gasoline used is linearly related to the number of miles driven. Unless discounts are given for buying large quantities of an item, the cost of a purchase of a number of the same item is linearly related to the number of items bought. The number of cups of coffee used at a church dinner is linearly related to the number of people in attendance.

Exercises 7.4

1. Use a rotation of a line about one of its points with an axis of rotation different from the line itself.
5. A parabola.

Exercises 7.5

1. (a) $(x - 1)^2 + (y - 5)^2 = 36$.
2. (a) $(-1, 1)$; 4.
3. $y^2 + 6y - 4x = -13$.
4. (a) An hyperbola opening right and left; cuts the x-axis at $(2, 0)$ and $(-2, 0)$.
   (c) Parabola opening to the right and passing through the origin.
   (e) An ellipse center at the origin.
5. It opens upward. It opens downward. It opens to the right. It opens to the left.

Exercises 8.1

1. The negation box gives as output the negation of the input.
   (a) $2 + 2 \neq 5$.
   (b) $2 + 3 \neq 6 - 1$.
   (c) $\sim(\sim p \vee \sim q) \longleftrightarrow \sim(\sim p) \wedge \sim(\sim q) \longleftrightarrow p \wedge q$.
3. (a) There are infinitely many simple closed curves that are topologically equivalent to a given simple closed curve. In fact, each such curve is topologically equivalent to every other one.
   (c) Most women have loved more than one man.

4. Yes; for each input there is exactly one output.
5. (a) The set of all nonvertical lines. Vertical lines have no slopes.
   (b) The conic sections that are circles or ellipses. The other conics do not enclose areas.

Exercises 8.2

1. Define a function f on the set of all polygons as follows: If P is a polygon, then $f(P)$ is the real number which is the number of sides of the polygon P. Define a function f on the set of circles as follows: If C is a circle, then $f(C)$ is the circumference of C.
3. x is a real number greater than or equal to 2, but at the same time is less than 3.
5. $(-2, -8)$, $(3, -3)$, $(4, -2)$, $(6, 0)$, $(7, 1)$, $(9, 3)$, $(12, 6)$, $(25, 9)$, $(100, 94)$, and $(2006, 2000)$.
6. (a) True.
   (c) False.
   (e) True.
7. (a) Neither 0 nor any negative number is the perimeter of a polygon.
   (b) A square of side 1, and octagon of side $\frac{1}{2}$, a hexagon of side $\frac{2}{3}$.
8. (a) $f(x) = x$.
   (c) $f(\text{polygon}) = $ number of sides of the polygon.

Exercises 8.3

1. (a)   (b)   (c)

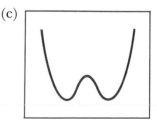

    The highest power of x indicates the number of times the curve changes direction from increasing to decreasing or from decreasing to increasing.
2. See the figure at the top of p. 340.
3. The curve on the right. The function is undefined for $x = 1$.
4. The graph crosses neither the x- or the y-axis. It consists of two disjoint pieces.
5. (a) The graph is called a step function, for it looks like a series of steps.
   (b) The graph is a "V" whose bottom is located at the origin.

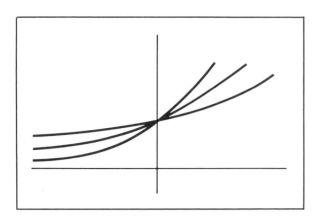

Exercises 8.4

1. Develop the formula $A(x) = 50x - x^2$. This function is maximum at $x = 25$ (study the graph), so $A(25) = 625$ square inches is the maximum value.
2. Let $w =$ the length of the side of the square cross section and let $V(w)$ represent the volume of the box. (Clearly, the volume will depend upon the value of w.) Then $V(w) = 4w^2(21 - w)$. This is minimum when $w = 14$. So the dimensions should be 14 by 14 by 25 inches.
3. Let ℓ denote the length of the base of the box. Then if $A(\ell)$ is the surface area of the box, $A(\ell) = 2\ell^2 + 256/\ell$. This is least when $\ell = 4$. Thus the minimal surface occurs when the box is a cube.
4. $A(x) = x[2 + 32/(x - 1)]$, where x is the width of a page and $A(x)$ is the area of a page. $A(x)$ is least when $x = 5$ inches. The length is then 10 inches.

Exercises 8.5

1. $\alpha = 45°$.
2. If we rounded the values of the angles $87°$, $88°$, and $89°$, it would appear that they were all equal, which they are not.
5. The proof follows by using the Pythagorean Theorem and the right triangle in Figure 8.13.
6. Sin $(90° - \alpha) = \cos \alpha$, since the graph of the cosine function is the graph of the sine function shifted to the left by $90°$.

Exercises 8.6

1. If α is the angle the wire makes with the top of the tower (draw a picture) then Cos $\alpha = \frac{1676}{2000} = .838$, approximately. So $\alpha = 33°$, approximately.

2. The tree is about 100 feet high.
4. The angle is about 76°.
5. About 68°.
6. About 5°.

Exercises 9.1

1. (a) Possible outcomes number 36 in all. Favorable outcomes: $1 + 1$, $1 + 2$, $2 + 1$, $1 + 3$, $3 + 1$, $1 + 4$, $4 + 1$, $2 + 2$, $2 + 3$, and $3 + 2$. $p = \frac{14}{36}$.
   (b) Possible: 36. Favorable: $1 + 1$, $2 + 2$, $2 + 4$, $4 + 2$, $3 + 3$, $3 + 5$, $5 + 3$, and 7 more. $p = \frac{16}{36}$.
   (c) Possible: 36. Favorable: $1 + 1$, $1 + 3$, $3 + 1$, $2 + 2$. $p = \frac{4}{36}$.
   (d) The only way to roll a number less than 3 is to roll $1 + 1$, which is even. So the only favorable outcomes are those listed in (b) above. (The point is not to count the outcome $1 + 1$ twice.) $p = \frac{16}{36}$.
2. (a) You will win on any of 18 numbers. There are 38 numbers altogether. $p = 18/38$.
   (b) $12/38$.
   (c) $12/38$. The numbers 0 and 00 do not belong to any column.
   (d) They are all equal, since there are equal numbers of reds, blacks, evens, and odds.
   (e) You win if any one of the five numbers comes up; so the probability of winning is $5/38$.
   (f) $4/38$.
3. There are $13 + 3 + 3 + 3$ cards that are hearts or face cards. Hence $p = \frac{22}{52}$. (You must be careful not to count the J$\heartsuit$, Q$\heartsuit$, and K$\heartsuit$ twice.)
4. The probability of 10's coming up on an American wheel is $\frac{1}{38}$. On a European wheel (with one fewer green numbers) the probability is $\frac{1}{37}$. Since $\frac{1}{37} > \frac{1}{38}$, you should prefer the European wheel (assuming, of course, that the payoffs are the same).
5. (a) An outcome is an individual card. The 52 outcomes are mutually exclusive and equally likely.
   (b) The outcomes are the women that a man might marry. These different outcomes are not equally likely, since a man is hardly as likely to choose a wife living on the other side of the earth as he is to choose one living in his own city. The outcomes, assuming no bigamy, are mutually exclusive.
   (c) The outcomes may be represented by this list: $25P$, $20P - 1N$, $15P - 2N$, $10P - 3N$, $5P - 4N$, $5N$, $15P - 1D$, $10P - 1D - 1N$, $5P - 1D - 2N$, $5P - 2D$, $3N - 1D$, and $1N - 2D$. These are mutually exclusive. They are probably not equally likely in actual practice. $2D - 1N$ is probably more likely than $25P$.
6. The favorable outcomes would be 2$\clubsuit$, 2$\diamond$, 2$\spadesuit$, K$\clubsuit$, K$\heartsuit$, K$\spadesuit$, J$\clubsuit$, J$\diamond$, J$\heartsuit$, 9$\diamond$, 9$\heartsuit$, and 9$\spadesuit$. There are $52 - 4 = 48$ cards from which to deal. $p = \frac{12}{48}$.

7. (a) You need any one of the four 2's. There are 48 cards from which to deal. $p = \frac{4}{48}$.
   (b) You need a 5, of which there are 4 in the 48 cards that remain. $p = \frac{5}{48}$.
   (c) You need either a 6 or an ace. There are 4 of each left in the 48-card deck. $p = \frac{8}{48}$.
8. Yes; the die can still come up only one number. Now 1 and 6 will be more likely to come up. Such crooked dice are used to throw 7's.
9. (a) Possible: B-B, B-G, G-B (in order of birth). Favorable: B-B. $p = \frac{1}{3}$.
   (b) Possible: B-B, B-G. Favorable: B-B. $p = \frac{1}{2}$.
10. All possible hands: K-Q, K-6, K-7, Q-6, Q-7, and 6-7. Ethelred has one of the hands K-Q, K-6, K-7, Q-6, and Q-7. The probability that he has two face cards is $\frac{1}{5}$. Ethelbert has one of the hands K-Q, K-6, and K-7. His probability is $\frac{1}{3}$. So Ethelbert is the more likely.

Exercises 9.2

1. The possible outcomes in tossing a head are "head" and "not head" and they have equal probability of occurring. Hence the probability of tossing a head is $\frac{1}{2}$. But on the European wheel the outcomes "red" and "not red" are not equally likely. The probability of "red" is $\frac{18}{37}$, while the probability of "not red" is $\frac{19}{37}$. The difference is, therefore, $\frac{1}{74}$.

Exercises 9.3

1. $C(52, 13)$.
2. $C(37, 3) \cdot C(22, 3) \cdot C(21, 3)$.
3. 26^3. $(21)(21)(5) + (21)(5)(21) + (5)(21)(21) + (21)(5)(5) + (5)(21)(5) + (5)(5)(21) + (5)(5)(5) = 8315$. $8315 - 125 = 8190$.
4. $(a + 1)(b + 1)(c + 1)(d + 1)(e + 1)$. $3300 = 2^2 \cdot 3^1 \cdot 5^2 \cdot 11^1$, so it has 36 divisors.
5. $C(7, 1)$. $C(7, 2)$. $C(7, 3)$. The sum equals $35 + 21 + 7 = 63$.
6. Possible plates: K-letter-letter-1-7-digit; K-letter-letter-digit-1-7; letter-K-letter-1-7-digit; and three more. Of the first type there are $(1)(26)(26)(1)(1)(10)$ possible; of the second $(1)(26)(26)(10)(1)(1)$; etc. There are six types of plates and $(26)(26)(10)$ possible of each type: 40,560 in total.
7. (a) A perfect square has the prime-factored form

$$p_1^{2k_1} p_2^{2k_2} \cdots p_s^{2k_s},$$

   where the p_i's are primes and the k_i's are nonnegative integers. By Exercise 4, such a number has $(2k_1 + 1)(2k_2 + 1) \cdots (2k_s + 1)$ divisors. This product is a product of odd numbers and hence is itself odd.
   (b) If the number is not a perfect square, then when written in prime factored

form at least one of the exponents will be odd. Then the number that expresses the total number of divisors will have an even factor and so will itself be even.

8. $C(6, 3) = 20$. $C(6, 3) - 4 = 16$. $2C(5, 3) - C(4, 3) = 16$.

Exercises 9.4

1. $C(13, 5)/C(52, 5)$. $C(9, 5)/C(42, 5)$.
2. $(\frac{1}{52})(\frac{1}{51})(\frac{1}{50})(\frac{1}{49})$. $1/C(52, 4)$.
3. $C(52, 2)$. $(\frac{13}{52})(\frac{13}{51})$. $(\frac{13}{52})(\frac{13}{51})(\frac{13}{50})$. $(\frac{13}{52})(\frac{13}{51})(\frac{12}{50})$.
4. $C(79, 19)/C(80, 20)$. $C(78, 19)/C(80, 20)$. $1 - C(78, 19)/C(80, 20)$.
5. $13[48/C(52, 5)] = \frac{1}{4165}$. Probability of a full house is about $\frac{1}{694}$. So a "four of a kind" is about six times as rare as a "full house."
6. $4/C(52, 13)$.
7. $(\frac{18}{37})^{32}$.
8. The total number of outcomes is $2^5 = 32$.
   (a) $\frac{1}{32}$.
   (b) $1 - \frac{1}{32}$.
   (c) $\frac{31}{32}$.
   (d) The probability of getting all heads or all tails is $\frac{1}{32} + \frac{1}{32} = \frac{2}{32}$. Thus the probability of at least one head and at least one tail is $1 - \frac{2}{32} = \frac{30}{32}$.
9. (a) $(52 - 13)/52 \cdot (51 - 13)/51 = \frac{741}{1326}$.
   (b) $1 - \frac{741}{1326} = \frac{585}{1326}$ for each suit.
10. $\frac{4}{52} \cdot \frac{3}{51} \cdot \frac{2}{50} \cdot \frac{1}{49}$.
11. A: $\frac{1}{4}$. B: $\frac{1}{8}$. C: $\frac{1}{8}$. D: $\frac{1}{8}$. E: $\frac{1}{8}$. F: $\frac{1}{4}$.
12. (a) $\frac{5}{14}$, $\frac{3}{14}$, $\frac{1}{14}$.
    (b) $\frac{4}{14}$, $\frac{6}{14}$, $\frac{3}{14}$.
13. $\frac{55}{100} \cdot \frac{6}{10} = \frac{33}{100}$.
14. The probability of the first player's dying is $\frac{1}{6}$. The probability that the first player does not die is $\frac{5}{6}$. If it is assumed that the first player does not die, the probability that second player will die is $\frac{1}{6}$. So the probability that the first player does not die, while the second player does, is $\frac{5}{6} \cdot \frac{1}{6} = \frac{5}{36}$, which is less than $\frac{1}{6}$. It would be better to go second, since the second player has a smaller probability of dying.
15. (Use analysis similar to Exercise 14.) The probability that the first player dies is $\frac{1}{6}$. The probability that the first player lives and the second player dies is $\frac{5}{6} \cdot \frac{1}{5} = \frac{1}{6}$. Hence it doesn't make any difference who goes first, from the point of view of probability.
16. The probability that the first player dies is $\frac{2}{6}$. The probability that the second player will die is $\frac{8}{30}$, which is less than $\frac{2}{6}$. Second. If the rules are changed so that the second player spins the cylinder, then the probability changes to $\frac{4}{6} \cdot \frac{2}{6} = \frac{8}{36} < \frac{2}{6}$, so it would be better to go second.
17. The probability of not hitting the bull's eye on the first throw is $\frac{2}{3}$. The same for the second the third throws. Therefore, the probability of not hitting the

bull's eye on any throw is $(\frac{2}{3})(\frac{2}{3})(\frac{2}{3})$. Hence the probability that he will hit the bull's eye at least once is $1 - (\frac{2}{3})^3$.

Exercises 9.5

1. $(\frac{5}{38})(+6) + (\frac{33}{38})(-1) = \frac{-3}{38}$. It is worse.
2. $(\frac{12}{38})(+2) + (\frac{26}{38})(-1) = \frac{-2}{38}$.
3. $(\frac{12}{38})(+2) + (\frac{26}{38})(-1) = \frac{-2}{38}$.
4. Let $x =$ number sold. $ME = (1/x)(500) + [(x - 1)/x](-1)$. A fair game has expectation 0, so solve the equation $ME = 0$ for x and get $x = 501$.
5. $ME = (\frac{1}{6})(6) + (\frac{1}{6})(5) + (\frac{1}{6})(4) + (\frac{1}{6})(3) + (\frac{1}{6})(2) + (\frac{1}{6})(1) = \frac{21}{6}$. You should expect to pay $\$\frac{21}{6}$ if the game is fair, more if it is not.
6. $ME = (3)(\frac{1}{6})(\frac{5}{6})(\frac{5}{6})(1) + (3)(\frac{1}{6})(\frac{1}{6})(\frac{5}{6})(2) + (\frac{1}{6})(\frac{1}{6})(\frac{1}{6})(3) + (\frac{5}{6})(\frac{5}{6})(\frac{5}{6})(-1) = \frac{-17}{216}$.

   So you should expect to lose a bit less than 8 cents each time you play.
7. No matter what were to happen, you would still lose \$3 each time you did this. For example, if three different numbers come up on the three dice, you would win \$1 for each of these numbers, but you would have had to pay \$6 for the privilege of playing. If the same number came up on all three dice, you would win \$3, but again, you would have had to pay \$6 in order to play.
8. First compute the probability that each of these events will occur: \$1-\$1 (probability $= (\frac{4}{8})(\frac{3}{7}) = \frac{12}{56})$, \$1-\$5 ($p = \frac{16}{56}$), \$5-\$5 ($p = \frac{2}{56}$), \$1-\$10 ($p = \frac{16}{56}$), \$5-\$10 ($p = \frac{8}{56}$), and \$10-\$10 ($p = \frac{2}{56}$). Then $ME = (\frac{12}{56})(2 - 6) + (\frac{16}{56})(6 - 6) + (\frac{2}{56})(10 - 6) + (\frac{16}{56})(11 - 6) + (\frac{8}{56})(15 - 6) + (\frac{2}{56})(20 - 6) = \frac{120}{56}$. She should agree; the probability is that she will get $\$\frac{120}{56}$ more than \$6.
9. At 40 per cent: ME of disarming $= (\frac{40}{100})(-10) + (\frac{60}{100})(10) = \frac{200}{100}$. ME of arming $= (\frac{40}{100})(5) + (\frac{60}{100})(2) = \frac{320}{100}$. Therefore, it is more desirable to arm than to disarm. Computing the ME's at the 30 per cent level, you get $\frac{400}{100}$ and $\frac{290}{100}$, in which case disarming is best.

Exercises 10.1

1. Cost of examining each and every member of the population would usually be prohibitive. The time required to examine each member of the population would make costs skyrocket. Testing may actually damage or destroy the product. Destroying a few light bulbs may be acceptable, but destroying them all doesn't make any sense at all unless it is the production technique itself that you are interested in. In many instances, the entire population will be inaccessible.
2. (a) Movie-goers may be less likely to be at home at this time.
   (b) People in this block are more likely to be coming from or going to the theater than people in other blocks.

(c) The movie-going habits of high school students might not be representative of the entire population. Also, school students may tend to attend more movies.

(d) This method involves only residents with telephones and, at best, only selected members of families with telephones.

3. The last census missed large numbers of inner-city and migrant families. The census has been greatly criticized as incorrectly representing poor people as a result.

Exercises 10.2

2. The frequency table is:

Datum	1	2	3	4	5	6	7	8	9	10	11	12	13
Frequency	2	4	4	6	3	6	7	5	4	8	4	6	6

3. (a) The frequency table will have a total of 38 weight groups: 95–96, 97–98, 99–100, 101–102, ..., 167–168, and 169–170. The corresponding frequencies are 3, 0, 1, 1, 1, 0, 0, 1, 0, 2, 0, 5, 3, 1, 0, 3, 1, 0, 1, 0, 0, 3, 0, 0, 0, 1, 1, 0, 0, 0, 0, 1, 0, 0, 2, 0, 0, and 1.

(b) There are 16 weight groups: 91–95, 96–100, 101–105, ..., 156–160, 161–165, and 166–170. The corresponding frequencies are 2, 2, 2, 1, 2, 8, 2, 3, 1, 3, 0, 2, 0, 1, 2, and 1.

(c) There are 8 weight groups: 91–100, 101–110, ..., 151–160, and 161–170. The corresponding frequencies are 4, 3, 10, 5, 4, 2, 1, and 3. Notice that the smoothed-out histograms more nearly resemble a normal curve as the size of the groupings increases.

4. You want to make it appear that the sales are soaring, so make the scale on the vertical axis such that the distance between, say, 1.200 and 1.250 is large. Also, make the width of the vertical bars small.

5. Make the scale on the vertical axis such that, for example, the distance between 1.200 and 1.250 is very small. Also, make the width of the vertical bars greater.

7. The frequency table is

Number of heads	0	1	2	3	4
Frequency	2	8	12	8	2

Exercises 10.3

1. | Exercise | Mean | Median | Range |
|---|---|---|---|
| (a) | 65.75 | 66.5 | 18 |
| (b) | 18.5 | 24 | 33 |
| (c) | $\frac{4}{7}$ | $\frac{1}{2}$ | 10 |
| (d) | 5 | 5 | 8 |
| (e) | 18 | 18 | 2 |
| (f) | 18 | 18 | 4 |
| (g) | x | x | 4 |

3. (a) The mode is E. The median is between C and D.
   (b) The mode is D; the median is between C and D.

6. 2760. The mode would tell them if the greatest number of people drank more or less coffee than the mean. For example, if the mode were less than the mean, this might imply that a few heavy coffee drinkers were increasing the mean. The airline might then save money by placing a limit on the number of cups a traveler could have. If the median differs greatly from the mean, this could imply that a number of heavy coffee drinkers (or nondrinkers) had a great effect on the mean.

7. The mean is 17 and the deviations from this mean are (in order) -2, 1, 4, -3, -5, 7, 1, -1, 0, -3, -4, -2, and 7. The sum of these deviations is 0. This is a general occurrence. The sum of the deviations from the mean will always be 0. This fact reveals an important property of the mean.

8. Let the score of the fourth examination be x. Then the average of 71, 77, 81, and x must equal 80. From this you obtain an equation in x that you can solve. $x = 91$.

9. 76.

10. 104. Yes. The mean of a group of numbers equals any constant plus the mean of the group formed by subtracting that constant from each number.

11. (a) 2141.
    (b) $x + 4$.

12. For example, (3, 4, 5, 6, 7), (2, 3, 4, 5, 11), (0, 0, 0, 0, 25), and (10, 1, 2, 3, 9).

Exercises 10.4

Exercise	Mean	Squared Deviations	Standard Deviation
(a)	6	1, 1, 1, 1, 0, 0, 0, 16	$\sqrt{\frac{20}{7}}$ or about 1.7
(b)	4	9, 4, 1, 0, 1, 4, 9	$\sqrt{\frac{28}{6}}$ or about 2.2
(c)	10	36, 16, 4, 0, 0, 0, 4, 16, 36	$\sqrt{\frac{112}{8}}$ or about 1.2
(d)	4	4, 4, 4, 0, 0, 0, 4, 4, 4	$\sqrt{\frac{24}{8}}$ or about 2.2
(e)	6.5	$\frac{1}{4}, \frac{1}{4}, \frac{1}{4}, \frac{1}{4}$	$\sqrt{\frac{1}{3}}$ or about .6
(f)	7.5	$\frac{1}{4}, \frac{1}{4}, \frac{1}{4}, \frac{1}{4}$	$\sqrt{\frac{1}{3}}$ or about .6
(g)	0	4, 1, 9, 4, 16, 9, 0, 1, 4, 4	$\sqrt{\frac{52}{9}}$ or about 7.5
(h)	4	.09, .16, .04, 0, .01, .01, .81	$\sqrt{\frac{1.12}{6}}$ or about 1.4

3. The mean is 8. The deviations squared are (in order) 16, 12.25, 9, 6.25, 4, 2.25, 1, .36, .16, .09, .04, .01, 0, .01, .04, .25, 1, 2.25, 4, 6.25, 9, 12.25, and 16. The sum of these values is 102.46. The standard deviation is $\sqrt{4.657}$ or about 2.16.

4. For example, 5, 5, 5, 5, and 5. A standard deviation of 0 means that all of the data are equal to the mean.

5. Mean A = 150 = Mean B. The standard deviation of scale A is about 0.4, while the standard deviation of scale B is about 0.3. Choosing the scale with the smaller standard deviation, we would want to select scale *B*.

6. Instructor A has received faculty ratings that are generally closer to the mean than has the other instructor. You would probably want to select instructor A.

7. The one with the smaller standard deviation is more reliable.

8. The sum of the deviations is always 0. Therefore, if they were not squared, the standard deviation would also be 0 and so would not reveal anything about the way the data is distributed about the mean.

9. It would be greater. By spreading the data out away from the mean the standard deviation would be increased.

Exercises 10.5

1. (a) 50 per cent. 0.5.
   (b) 5.48 per cent. 94.52 per cent. 0.0548. 0.9452.
   (c) 91.85 per cent. 0.9185.
   (d) 9.61 per cent.
   (e) 0.0633.
2. $\mu - 1.28\sigma$. $\mu + 1.28\sigma$.
3. (a) 101.65.
   (b) 98.52.
   (c) 99.33.

4. (a) 4.75 per cent. 0.0475. 0.9525.
   (b) 99.38 per cent. 0.62 per cent.
   (c) 40.66. 59.34.
   (d) The 5.2-pound baby.
5. (a) $x = 41.2$.
   (b) 3.07 per cent. 3.07 per cent. 0.03. 0.03.
   (c) 50,000. 50,000.
6. (a) 550.
   (b) 192.
7. (a) 520. 520. 520.
   (b) $\sigma = 45/.47$ or about 96.
   (c) 0.62 per cent.
   (d) 655.
8. (a) Negligible.
   (b) Yes.
   (c) Yes.
9. (a) 2.28 per cent. 13.59 per cent. 68.26 per cent. 13.59 per cent. 2.28 per cent.
   (b) None.
10. (a) 90.
    (b) 80.

Exercises 10.6

1.

0	0		30	20		84	56		84.7	30.3
100	0		30	20		36	24		180.3	64.7

2. (a) The property of being female is independent of the property of habitually using perfume.
   (b) The property of being married is independent of the property of owning your own home.
   (c) The property of being good at arithmetic computation is independent of the property of being a good accountant.

3.

30 (30)	20 (20)	50
90 (90)	60 (60)	150
120	80	200

The key number is 0, so we should accept the hypothesis of independence.

4.

45 (40)	35 (40)	80
65 (60)	55 (60)	120
100	100	200

Key number $= \frac{25}{40} + \frac{25}{40} + \frac{25}{60} + \frac{25}{60} = 2.08$, about. Hence accept independence for critical numbers of either 3.84 or 6.63.

5. (a)

86	29
94	31

(b) Yes.

6. (a)

68	37
33	17
29	16

(b) Key number $= \frac{9}{68} + \frac{9}{37} + \frac{49}{33} + \frac{49}{17} + \frac{16}{29} + \frac{16}{16}$ or about 6.3.

(c) Accept independence at the 99 per cent level. Reject independence at the 95 per cent level.

7.

18	16	16
24	23	23
98	91	91

Key number $= (17)^2/18 + (11)^2/16 + (6)^2/16 + (11)^2/24 + (8)^2/23 + (3)^2/23 + (28)^2/98 + (19)^2/91 + (9)^2/91$, or about 47. We may reject independence at either level.

Index

Symbolic logic, 24
Symmetric property, 49

T

Tautology, 33, 36
Terminating decimal numeral, 180
Thales of Miletus, 15
Theorem, 17
Topologically equivalent, 146
Torus, 130
Transitive property, 49
Translation, 133
Truth, 23
  tables, 27
Twin primes, 89

U

Uncountable cardinal number, 123
Undefined concept, 18
Union of sets, 109
Unique Parallel Postulate, 20
Universal set, 102
Universal statement, 39

V

Validity, 23
Venn diagram, 111
Vinogradoff, 89

W

Whole number, 44
  prime factorization of, 91
  square free, 92
Whole numbers, addition of, 71
  division of, 76
  equality of, 48
  greatest common divisor of, 93
  least common multiple of, 94
  multiplication of, 71
  relatively prime, 94
  subtraction of, 75

Z

Zero, as annihilator, 81
  division by, 58
  as exponent, 85
  history of, 81